PRINCIPLES OF OPTICAL CRYSTALLOGRAPHY

A. V. SHUBNIKOV

TRANSLATED FROM RUSSIAN

Springer Science+Business Media, LLC 1960

The original Russian text was published for the Institute of Crystallography by the Academy of Sciences USSR Press in Moscow in 1958.

Library of Congress Catalog Card Number 59-14222

ISBN 978-1-4899-4883-0   ISBN 978-1-4899-4881-6 (eBook)
DOI 10.1007/978-1-4899-4881-6

CONTENTS

CONTENTS (Continued)

CONTENTS (Continued)

CONTENTS (Continued)

FOREWORD

The present book is devoted to the branch of crystallography which is concerned with optical properties of crystals. It follows the course on optical crystallography given in the Physics Faculty of the Moscow State University. This explains the absence of isolated details and even whole branches of the subject, for example, reflection of light from crystal plates, changes in optical properties of crystals under the action of external effects, etc. The book also does not include descriptions of apparatus and experimental methods which are covered by practical classes running in parallel with the above theoretical course. In the preparation of the present edition of the "Principles of Optical Crystallography," much of the material in the earlier version has been rewritten, various criticisms have been taken into account, and some errors detected by the author himself have been rectified.

The author wishes to express his gratitude to N. M. Melankholin for his careful reading of the manuscript and a number of valuable suggestions which were incorporated in the final draft.

The Author

THE SUBJECT OF OPTICAL CRYSTALLOGRAPHY

Optical crystallography deals with the optical properties of crystals and, being a part of physical crystallography, it is concerned with the classification of crystals according to their optical properties and the connection between optical and other properties of crystals, in particular, their symmetry.

Optical crystallography has naturally much in common with crystal optics, which is a branch of physics, and methods of measuring optical constants of crystals. The latter subject has been already treated in a large number of excellent texts and is therefore not considered in any detail in the present book.

BASIC IDEAS FROM THE OPTICS OF ISOTROPIC
TRANSPARENT MEDIA

The Laws of Reflection and Refraction. Light incident on a plane
and smooth boundary between two transparent isotropic media is reflected so that (1) the incident ray, the reflected ray, and the normal
to the reflecting surface all lie in the same plane, and (2) the angle
of incidence i_1 and the angle of reflection i_2 are equal.

When light strikes a plane boundary between two transparent isotropic media it is refracted so that (1) the incident ray, the refracted
ray, and the normal to the refracting surface all lie in the same plane,
and (2) the ratio of the sine of the angle of incidence to the sine of the
angle of refraction is a constant independent of the angle of incidence
and is given by

$$\frac{\sin i_1}{\sin i_2} = n_{12} .$$ (1)

The quantity n_{12} is known as the refractive index of medium 2 relative to medium 1. When the rays are reversed, we have

$$\frac{\sin i_2}{\sin i_1} = \frac{1}{n_{12}} = n_{21}.$$ (2)

When a refractive index is quoted for a given medium without spec-
ifying a second medium, the refractive index is understood to be relative to vacuum, i.e., it is assumed that light enters the given medium
from a vacuum. The refractive index is then denoted by the letter n
without subscripts.

In the case of ideally transparent media, the refractive index is
an essentially positive quantity greater than unity. The relative refractive index can, of course, be less than unity. By definition, the
refractive index of vacuum is unity and the refractive index of air is
very close to unity.

5

If a ray of light passes successively through media 1, 2, 3, 4, the refractive index of the last medium relative to the first is given by

$$n_{14} = n_{12} \cdot n_{23} \cdot n_{34}. \tag{3}$$

Given two media, the medium with the larger refractive index is said to be the more refracting one. On passing from a less refractive medium to a more refracting one, a ray of light should, according to the laws of refraction, be bent toward the normal, and when it travels in the opposite direction it should be bent away from the normal.

The phenomenon of refraction was known in antiquity but was first formulated in the above form by Snell in 1618 and, independently, by Descartes in 1637.

Total Internal Reflection. Suppose a ray of light passes from a more refracting medium 2 to a less refracting medium 1 (Fig. 1). From the law of refraction we have

$$\frac{\sin i_1}{\sin i_2} = n_{12} > 1. \tag{4}$$

Let us now gradually increase i_2 until it reaches the maximum value I_2 allowed by this formula. When i_2 is equal to this maximum value, the angle i_1 will be equal to 90° and its sine will be unity. It follows that

$$\frac{1}{\sin I_2} = n_{12}. \tag{5}$$

If the angle of incidence is greater than I_2, then $\sin i_2$ should be greater than unity, and this is impossible. It means that for angles

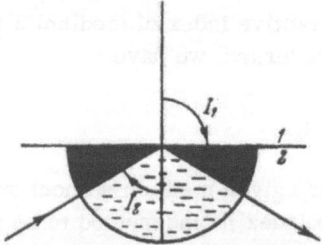

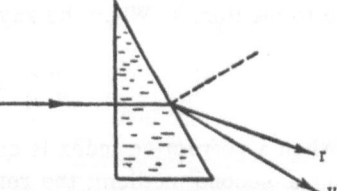

Fig. 1. Total internal reflection of light. Total internal reflection takes place when the angle of incidence I_2 of a ray incident from a more refracting medium 2 (glass) on a less refractive medium 1 (air) reaches at least such a value that the angle of incidence and the refractive index are related by $n = 1/\sin I_2$.

Fig. 2. Refraction of white light by a glass prism. The white light is decomposed into a spectrum (dispersion). In the case of normal dispersion the violet rays are refracted more strongly than the red rays.

6

of incidence greater than I_2, refraction is absent. Experiment shows that in this case the light is totally reflected. Such a reflection is known as total internal reflection. The angle I_2 is called the critical angle. For glass, whose refractive index relative to air is $n = 1.5$, the critical angle is equal to 41°48'. Refractive indeces are often determined by measuring I_2.

Dispersion. In 1672 Newton showed experimentally that a beam of white light passed through a glass prism is decomposed into a spectrum consisting of a large number of colors, from the red to the violet, which gradually merge into each other. It is convenient to divide the various colors of the spectrum into eight intervals: red, orange, yellow, yellow-green, green, light blue, dark blue and violet, and to arrange them in a color rosette as shown in Fig. 3. The decomposition of white light into a spectrum (dispersion) indicates that rays of different color have different refractive indices. In the case of normal dispersion which is observed in colorless media, red light has the lowest and violet light the highest refractive index. In the case of anomalous dispersion which is observed in colored media, this is not the case at the point in the spectrum where the absorption of light takes place.

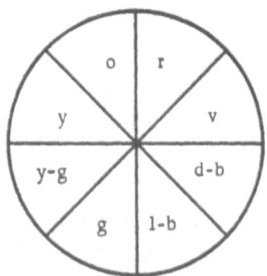

Fig. 3. An eight color rosette. Opposite sectors correspond to complementary colors. The colors are denoted by their first letters: r) red, o) orange, y) yellow, y-g) yellow-green, g) green, l-b) light blue, d-b) dark blue, v) violet.

Light of a given color is called monochromatic. Strictly monochromatic light cannot be obtained in practice since this would involve the extraction from a spectrum of an infinitely narrow line. The intensity of such a line would be infinitely small so that it could not be observed. Infinitely narrow rays and monochromatic light are only useful scientific abstractions. In practice, one deals with beams of light and mixed colors consisting of many, even though very close to each other, single colors.

Light Waves. In a uniform medium light is not propagated instantaneously but has a certain constant and finite velocity. This velocity was first determined by Römer in 1676 and was found to be $c = 3 \cdot 10^{10}$ cm/sec for both white and monochromatic light of any color.

Experiment shows that there are only two kinds of uniform rectilinear motion in nature, namely, inertial motion and wave motion.

7

Since the motion of light in a uniform medium is both uniform and rectilinear, it appears that there are only two possible assumptions as to its nature. The first of these is ascribed to Newton (1672) who looked upon light as a stream of material particles, and the second to Huygens who thought that light has a wavelike nature. We know that light may be looked upon both as a stream of uniformly moving particles and as uniformly moving waves. Classical crystal optics and optical crystallography are based on the wave theory of light.

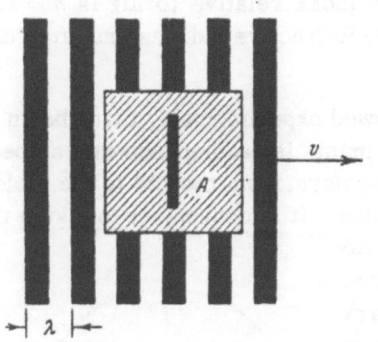

Fig. 4. Moving linear or plane waves. The wavelength is equal to twice the width of a white or black band.

Simple Wave Model. Plane waves may be characterized, independently of their physical nature, by at least two quantities, namely, the velocity of propagation v and the wavelength λ. Thus a simple wave model may consist of a plane system of white and black bands of equal width and moving with a constant velocity v in the direction of the normal to the bands (Fig. 4). The wavelength λ will then be equal to the sum of the width of a black and a white band. If the number of waves passing per second in front of a slit A through which they are observed is ν, then the velocity v is given by

$$v = \nu\lambda.$$ (6)

At the same time, an observer looking through the slit will see vibrations, i.e., the slit will be alternately covered by black and white bands with a time interval of $\frac{T}{2}$. The quantity ν is known as the frequency of the vibrations and T is the period. These two quantities are inversely proportional to each other, the relation between them being

$$T = \frac{1}{\nu}.$$ (7)

It is convenient to introduce a quantity k which is the reciprocal of the wavelength, i.e.,

$$k = \frac{1}{\lambda}.$$ (8)

It is known as the wave number (wave density) and is equal to the number of waves per unit length along the direction of propagation. We

8

recall that a constant velocity v is equal to the distance traveled divided by the time taken, i.e.,

$$v = \frac{x}{t}. \tag{9}$$

We shall also use the reciprocal of the velocity

$$\frac{1}{v}, \tag{10}$$

which characterizes the "slowness" of propagation of the waves.

So far we have only considered traveling waves. When $v = 0$ traveling waves become stationary. They may be represented by a model consisting of a stationary system of black and white bands. Stationary waves must not be confused with standing waves. The latter waves may be represented by a system of black and white bands in which the bands periodically change from black to white.

We have described a model of linear waves. If we replace the black and white bands by similar layers we shall obtain a model of plane waves. This may then be easily generalized to spherical waves, a section through which is shown in Fig. 5, and to waves having arbitrary wave surfaces.

Fig. 5. A system of stationary circular or spherical waves.

As was pointed out above, our model is only a simple one. It does not take into account the physical nature and the intensity of the waves, nor the form of the vibration. It is equally convenient for both transverse and longitudinal waves, as well as for waves which are neither transverse or longitudinal, for example, temperature waves.

Refraction of Light in the Wave Theory. It is known that the wavelength of monochromatic light changes when it passes from one medium to another. The ratio of the two wavelengths λ_1 and λ_2 is equal to the relative refractive index, i.e.,

$$n_{12} = \frac{\lambda_1}{\lambda_2}.$$

A refractive index determined in this way should, of course, be independent of the angle of incidence since the wavelengths in the first and second media can only depend on the internal properties of these media and not on what happens on the boundary between them. Clearly, refraction essentially consists of a change in the wavelength and not a change in the direction in which the light was originally traveling. The change in wavelength is observed even when, according to the law

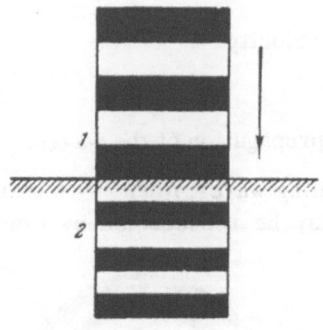

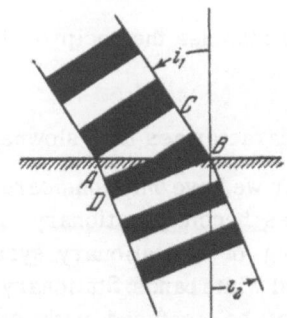

Fig. 6. Reduction in the wavelength of light on transmission from a less refracting medium 1 to a more refracting medium 2.

Fig. 7. Refraction of plane waves. The law of refraction is derived as a consequence of the reduction in wavelength.

of refraction, there should be no refraction at all, i.e., in the case of normal incidence (Fig. 6). This special case is in fact the most typical and important.

Figure 7 illustrates the refraction of waves in the case of oblique incidence on the boundary separating two media. From triangles ABC and ABD we have

$$\frac{CB}{AD} = \frac{\sin i_1}{\sin i_2} = \frac{\lambda_1}{\lambda_2} = n_{12}, \tag{11}$$

i.e., the Descartes-Snell law of refraction is a consequence of the assumption that there is a wavelength change when light passes from one medium to another.

Color in the Wave Theory. It is normally assumed that the color of monochromatic light does not change when it passes from one transparent medium to another. In the wave theory this is explained by saying that the frequency ν (and its reciprocal T) remains constant. It follows that, similarly to pitch in sound, color depends only on frequency.

In the above discussion optical quantities were denoted by letters, e.g., n, λ, k, v, etc., and the numerical indices were used to indicate the medium to which the given quantity refers. These indices clearly have no meaning in the case of ν and T since these two quantities are independent of the properties of the medium. Bearing this in mind, we can write down two equations for monochromatic light passing from one medium to another:

$$v_1 = \nu\lambda_1,$$
$$v_2 = \nu\lambda_2$$

or

$$\frac{v_1}{v_2} = \frac{\lambda_1}{\lambda_2} = n_{12}. \qquad (12)$$

It follows that when light passes through the boundary between two transparent media, both the wavelength and the velocity change by the same factor. This factor is equal to the relative refractive index.

In the case when light enters the medium from a vacuum, expression (12) is usually replaced by

$$n = \frac{c}{v} = \frac{\lambda_0}{\lambda}, \qquad (12')$$

where n is the refractive index of the medium, c and λ_0 are the velocity and wavelength in a vacuum, and λ is the wavelength in the medium.

It was pointed out above that the refractive index (relative to a vacuum) of transparent media is greater than unity. It follows that the velocity of light in a medium is less than in a vacuum, i.e.,

$$v < c. \qquad (13)$$

We know already that monochromatic waves of different frequencies are propagated in a vacuum with the same velocity. Hence, if one monochromatic beam of light has a wavelength λ' and a frequency ν' in a vacuum, and the corresponding quantities for a second beam are λ'' and ν'', then in accordance with expression (6) we have

$$\lambda'\nu' = c,$$
$$\lambda''\nu'' = c$$

or

$$\frac{\lambda'}{\lambda''} = \frac{\nu''}{\nu'}. \qquad (14)$$

The frequency of light vibrations corresponding to the yellow line emitted when sodium compounds are heated is of the order of

$$\nu_{Na} \approx 0.5 \cdot 10^{15}.$$

The wavelengths (in a vacuum) corresponding to the visible part of the spectrum occupy one octave between 400 mμ and 800 mμ (Table 1).

11

TABLE 1

Approximate Wavelengths of Monochromatic Rays
in a Vacuum

Color	λ, mμ	Color	λ, mμ
Ultraviolet	< 390	Yellow-green	550 – 575
Violet	390 – 450	Yellow	575 – 585
Light blue	450 – 480	Orange	585 – 620
Dark blue	480 – 510	Red	620 – 760
Green	510 – 550	Infrared	> 760

On the short wavelength side, the visible region of the spectrum lies next to the invisible ultraviolet waves, and on the long wavelength side it lies next to the invisible infrared waves.

Interference and Group Waves. By interference we shall understand (in the widest sense) the formation of complex waves from simple waves. These complex secondary waves will be called group waves. They have a group velocity, frequency, wavelength, etc.

The formation of group waves is best illustrated by the superposition of two systems of white and black bands at a small angle α to each other (Fig. 8). It is clear from Fig. 8 that group waves have

Fig. 8. Appearance of secondary (group) waves
as a result of the superposition of primary waves
of different wavelength and direction.

a larger wavelength than the primary waves and are, in general, differently oriented than the primary waves. If the primary waves move in the direction of their normals then, in general, the group waves will also move, although the group velocity may be different from the velocity of the primary waves both in direction and magnitude. The following special cases are of particular interest.

If the wavelengths and velocities of the primary waves are the same, the group waves will be propagated along the bisector of the obtuse angle between the primary waves, or the acute angle between

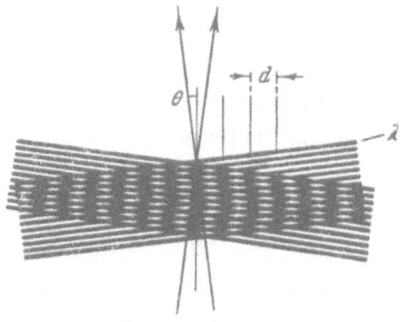

Fig. 9. Formation of group waves as a result of the superposition of primary waves of equal wavelength but different direction.

the directions of the primary wave velocities. It is easy to show that in this case the wavelength of the group waves L, the wavelength of the primary waves λ , and the angle α are related by the expression

$$\lambda = 2L \sin \frac{\alpha}{2} . \tag{15}$$

This formula is exactly the same (when L is replaced by d and α by 2θ, Fig. 9) as the Bragg-Wulff formula for the reflection of x-rays by a family of parallel planes in a crystal lattice, which in this case take the place of stationary group waves formed by the incident and reflected x-rays. The system of lattice planes itself can, clearly, be treated as a system of stationary waves.

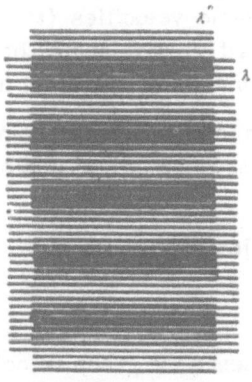

Fig. 10. Formation of group waves from primary waves of different wavelength but the same direction.

Let us now consider a second example of interference which is known as beats. Beats will occur when the interfering waves are parallel to each other and their wavelengths are different (Fig. 10). It is easy to show that in the case of beats the group wavelength is given by

$$L = \frac{\lambda'\lambda''}{\lambda' - \lambda''} , \tag{16}$$

while the density of the group waves is given by

$$K = k'' - k'$$ (17)

and the group velocity by

$$V = \frac{\lambda' v'' - \lambda'' v'}{\lambda' - \lambda''},$$ (18)

where λ', λ'' are the wavelengths, k', k'' are the wave numbers and v', v'' are the velocities of the primary waves.

Let us consider expression (18) in greater detail assuming that $\lambda' > \lambda''$. If both the velocities are positive, the numerator can still be either positive or negative. This means that the group waves can be propagated either in the direction of the primary waves or in the opposite direction. If v' and v'' have different signs, the group velocity may have the direction of either v' or v''. In a special case it may be zero and this will occur when

$$\lambda' v'' = \lambda'' v'.$$

Since λ', λ'' are positive this condition can only be satisfied if v' and v'' have the same signs. It follows that the group velocity can only have a nonzero value when both the primary waves move in the same direction.

The third special case of interference is the complete mutual compensation of the two systems of primary waves. This occurs when the primary systems are parallel, have the same wavelengths ($\lambda' = \lambda''$), the same velocities ($v' = v''$) and one of the systems leads the other by half a wavelength, or an odd number of wavelengths. The wavelength of group waves in this case is infinite.

In optics, monochromatic light waves play the role of primary waves. Since in the case of monochromatic waves in a vacuum

$$v' = v'' = c,$$

it follows from (18) that

$$V = c.$$ (19)

and hence the group velocity of light in a vacuum does not differ from the velocity of monochromatic waves, i.e., the so-called phase velocity (p. 16).

Harmonic Vibrations of the Light Vector. Consider a vector A (Fig. 11) which rotates uniformly about the origin O' so that its end point successively passes over the points $0, 1, 2, 3 \ldots$ on a circle. The

14

projection a of the vector A on one of the diameters of the circle is said to execute harmonic or sinusoidal vibrations.

When the vector A rotates uniformly, the phase angle φ is clearly proportional to time so that

$$\varphi = \omega t, \tag{20}$$

where ω is a constant known as the angular velocity or angular frequency. During the time T of one revolution (one vibration) the vector A describes an angle equal to 2π, so that

$$2\pi = \omega T,$$

and hence

$$\omega = \frac{2\pi}{T}, \tag{21}$$

and

$$\varphi = \frac{2\pi t}{T}. \tag{22}$$

The length of the projection of the vector A is given by

$$a = A \sin \varphi = A \sin \omega t = A \sin \frac{2\pi t}{T}. \tag{23}$$

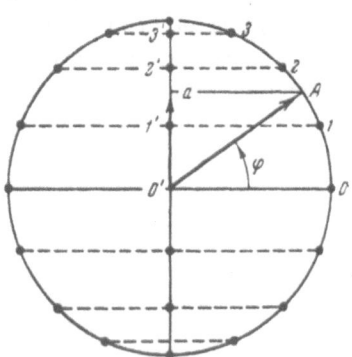

The maximum value of a, which is equal to A, is known as the amplitude of the vibrations and the quantity T is known as their period.

Fig. 11. When the vector A rotates uniformly its projection a on the vertical diameter of the circle executes harmonic vibrations with an amplitude equal to the length of the vector A.

When the end point of the vector A moves uniformly over the circle and describes the arcs 01, 12, 23 . . . in equal times, the end point of the vector a will move nonuniformly and will describe during the same times the unequal segments 0'1', 1'2', 2'3', . . . of the diameter of the circle. The end point of the vector a will have its maximum velocity when it passes through the center of the circle, and the velocity will be zero when the vector reaches its maximum value A. The general expression for this velocity can be obtained by differentiating (23) with respect to time:

$$v_a = A\omega \cos \omega t. \tag{24}$$

Equation (20) was obtained on the assumption that when $t = 0$ the rotating vector coincides with the horizontal diameter of the circle. If this is not the case, equation (20) may be written in the more general form:

15

$$a = A \sin(\omega t - \Delta) = A \sin(\tau - \Delta). \qquad (25)$$

In this case the vector A will pass through the horizontal diameter when

$$\tau = \Delta \quad \text{or} \quad t = \frac{\Delta}{\omega}.$$

Clearly, formulas (20), (21), (22) and (24) must then be similarly generalized.

In 1815 Fresnel suggested that the wave nature of monochromatic rays may be associated with harmonic vibrations of the particles of a hypothetical ether, that the vibrations are transverse relative to the direction of propagation, and that consequently the displacement a of an ether particle given by expression (23) can be identified with the light vector.

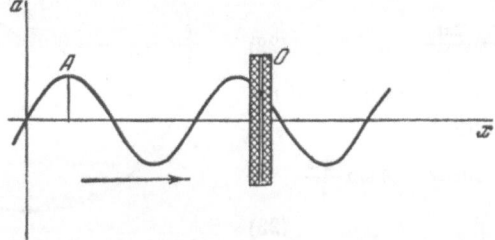

Fig. 12. A simple model of a ray with transverse vibrations. The sinusoid moves uniformly in the x direction. The point on the sinusoid seen through the slit O executes harmonic vibrations with an amplitude A.

Model for Transverse Sinusoidal Waves. A stationary sinusoid with an amplitude A (Fig. 12) which represents an instantaneous photograph ($t = $ const) of a moving wave may serve as a model of stationary transverse sinusoidal waves of any physical nature. If this sinusoid is brought into motion in the direction of the x axis with a velocity v_x, one obtains a model of a moving wave. If one looks at the sinusoid through the slit O, one sees a single point executing harmonic vibrations. Conversely, if one imagines that all the points which were originally on the x, axis execute transverse sinusoidal vibrations relative to it, so that each point lying to the right of the given point vibrates with a correspondingly lower phase, then the whole aggregate of vibrating points will be represented by a sinusoid moving with the velocity v_x. to the right. Any given phase of the vibrations will, clearly, move with this velocity. It follows that the velocity v_x is the phase velocity.

Equation for Sinusoidal Waves. If a particle in an elastic medium is brought into harmonic vibrations, the neighboring particles will also begin to vibrate but the phase of these vibrations will decrease with increasing distance from the source. It follows that

$$\varphi_0 - \varphi_x = \varkappa x, \tag{26}$$

where φ_x is the phase of a particle at a distance x, from the source, φ_0 is the phase at the source and $\varkappa$ is a constant. Clearly, two particles will be in phase ($\varphi_0 - \varphi_x = 2\pi$) if they are separated by one wavelength, i.e., $x = \lambda$. Hence

$$2\pi = \varkappa\lambda,$$
$$\varkappa = \frac{2\pi}{\lambda}.$$

Substituting this expression for $\varkappa$ into (26) we have

$$\varphi_x = \varphi_0 - \frac{2\pi x}{\lambda} = 2\pi\left(\frac{t}{T} - \frac{x}{\lambda}\right).$$

If we substitute this expression into (23) we obtain

$$a = A \sin\left[2\pi\left(\frac{t}{T} - \frac{x}{\lambda}\right)\right]. \tag{27}$$

This is the equation for sinusoidal waves which we set out to find. It may be written in various forms by introducing into it the quantities v, ω, ν, k, and T. Thus, substituting $\lambda = vT$ we have

$$a = A \sin\left[\frac{2\pi}{T}\left(t - \frac{x}{v}\right)\right]. \tag{28}$$

If we introduce the angular frequency $\omega = \frac{2\pi}{T}$ and the angular wave number $\varkappa$, the wave equation assumes the form

$$a = A \sin(\omega t - \varkappa x). \tag{29}$$

If we introduce the frequency $\nu = \frac{1}{T}$ and the wave number $k = \frac{1}{\lambda}$, we obtain yet another new form of this equation:

$$a = A \sin[2\pi(\nu t - kx)]. \tag{30}$$

Equations (25) – (30) were derived for sinusoidal waves of any nature, in particular, monochromatic light waves. We are interested in the following three special cases.

1. In the case of plane waves (parallel beam of rays) propagated in an ideally transparent nonabsorbing medium, the amplitude A remains constant for all values of x.

17

2. In the case of spherical waves, the energy, which is proportional to A^2, is inversely proportional to x^2, i.e., A is inversely proportional to x:

$$A = \frac{A_0}{x},$$

where A_0 is the initial amplitude.

3. In the case of plane waves propagated in an absorbing medium we shall find that (p. 184)

$$A = A_0 e^{-\frac{kx}{2}},$$

where k is the absorption coefficient.

Electromagnetic Theory of Light. According to Maxwell (1865), light waves are simply very short electromagnetic waves. In a vacuum, the light vector may be identified with the electric field E, while in a medium the corresponding quantity is the displacement D, the connection between the two quantities in the case of an isotropic medium being

$$D = \varepsilon E,$$

where ε is the dielectric constant of the medium.

Electromagnetic waves always appear in a dielectric if for some reason an electric field is suddenly produced in it and, similarly to a disturbance on a water surface, does not remain at the point but is propagated in the form of waves into the surrounding medium. The electrical waves give rise to magnetic waves which consist of changes in the magnetic field H (in a vacuum) or the magnetic induction $B = \mu H$ (in a medium), where μ is the magnetic permeability.

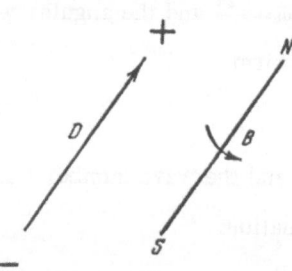

Fig. 13. The polar vector D (displacement) and the axial vector B (magnetic induction).

The electric vectors E, D are polar vectors and are usually represented by straight-line arrows. This representation of polar vectors reflects their $\infty \cdot m$ symmetry (infinite axis and longitudinal symmetry planes). The magnetic vectors H, B are axial vectors. They are also usually represented by straight-line arrows but this representation does not correctly reflect their true $\infty : m$ symmetry (infinite axis with a transverse symmetry plane and center).

18

Magnetic vectors are therefore best represented by a segment of a straight line with a circular arrow showing the direction of the Amperian currents in a magnet (Fig. 13). This means that electromagnetic waves may be represented by the model shown in Fig. 14. The

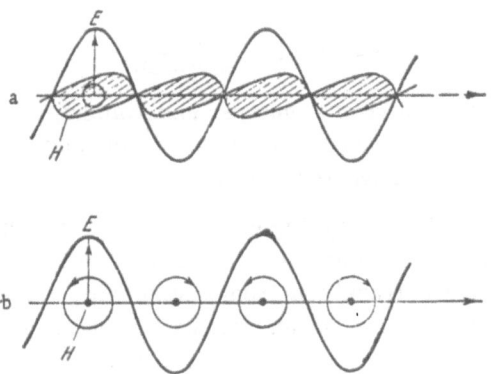

Fig. 14. Electromagnetic model of a beam of light showing correctly its symmetry. a) Three-dimensional representation of the model; b) projection on the plane of symmetry (plane of vibrations of the E vector).

model may easily be used to determine the direction of propagation of the waves (the direction of the Poynting vector) from the mutual orientation of the vectors E and H or D and B. Below, we shall use a simplified model of light waves which consists of a single moving "electric" sinusoid.

Cause of Dispersion of Light. The main result of the electromagnetic theory of light is Maxwell's law, which in the case of transparent isotropic dielectrics may be written in the form

$$n = \sqrt{\varepsilon}. \tag{31}$$

Maxwell assumed that ε is a constant quantity independent of frequency so that the refractive index n should also be independent of frequency, and this, as we know now, is not the case. Thus the electromagnetic theory of light in its original form could not explain the dispersion of light.

A successful explanation of this phenomenon was achieved only as a result of simultaneous application of the Maxwell theory and the Lorentz electron theory (1896). The theory of dispersion is based on the assumption that the alternating electric field of light waves passing

19

through a transparent body brings into oscillation electrons which are bound to the atoms by elastic forces. The vibrating electrons then emit secondary electromagnetic waves which interact with the incident waves. It finally turns out that ε may be represented by the following formula

$$\varepsilon = n^2 = 1 + \frac{4\pi N e^2}{m\,(\omega_0^2 - \omega^2)}, \qquad (32)$$

where N is the number of atoms per unit volume, e is the electronic charge, m is the electronic mass, ω_0 is the angular velocity of an electron and ω is the angular frequency of the incident waves.

Light Quanta (Photons). Light is always emitted (both visible and invisible) when the particles in a medium (atoms, molecules, electrons) experience a transition from a state with a higher energy to a state with a lower one, i.e., a transition from a higher energy level to a lower one. According to modern theories these energy levels are discrete, i.e., they are separated from each other by finite intervals, so that a transition from one energy state to another can only occur through a jump with the emission or absorption of a finite portion of energy, i.e., a quantum.

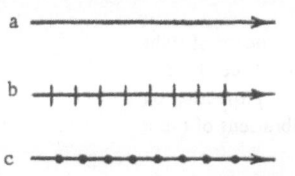

Fig.15.Conventional representation of rays of light. a) Unpolarized; b) polarized with vibrations in the plane of the paper; c) polarized with vibrations perpendicular to the plane of the paper.

According to Planck (1900) the quantum of energy is

$$E = h\nu_{ik}, \qquad (33)$$

where h is Planck's constant and ν_{ik} is the frequency of the emitted or absorbed light during a transition from the ith state to the kth state. It is assumed that the states are numbered in the order of increasing or decreasing energy.

It follows from this expression that all quanta of light are monochromatic, that the ultraviolet quantum is greater than the violet, that the violet quantum is greater than the red, and that the red is greater than the infrared. The simpler the internal structure of the emitting particle the smaller is the number of its energy levels. This explains why the spectrum of a rarified gas consisting of separate atoms weakly interacting with each other consists of a relatively small number of lines, while the spectrum of an incandescent solid body (a very large "particle") or a strongly compressed gas is continuous and consists of a very large number of lines close to each other.

According to Einstein (1905), the quantum of light energy, i.e., a photon, is equal to the product of the photon mass and the square of the velocity of light in a vacuum:

$$E = mc^2 = h\nu. \tag{34}$$

This means that a photon has both particle (mass) and wave (frequency) properties.

Let us, however, return to the wave model. According to the wave theory, a monochromatic beam of light may be looked upon as an infinite sinusoid moving with a constant velocity along its axis. According to the quantum theory this sinusoid may be looked upon as broken up into equal parts which are distributed irregularly along the beam but having a certain definite mean density. Each piece of the sinusoid is polarized, i.e., the vibrations are in a single plane which is known as the vibration plane. A beam of light may not be polarized as a whole since the different parts of the sinusoid may be polarized in different planes.

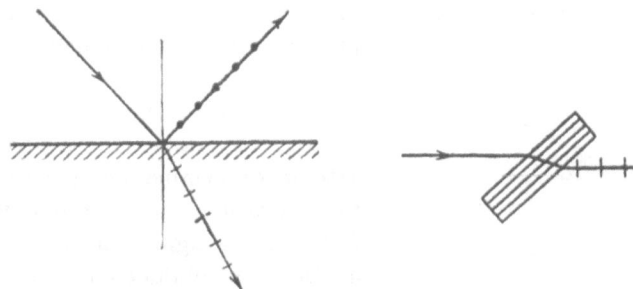

Fig. 16. Polarization of natural light on reflection and refraction.

Fig. 17. Polarization of light produced by a pile of glass plates.

If two beams of light of different origin are propagated along the same path they will interact (or, in a broad sense, interfere) with each other only at those points where the sections of the sinusoids will accidentally superimpose, while when sections of the one sinusoid fall into the gaps between the sections of the other, interference will not take place. Such beams cannot give a macroscopic interference effect, i.e., an effect which may be observed in the usual way. If on the other hand a beam is first split in some way into two beams of similar structure, and the path difference between the two beams is not too large, i.e., the corresponding paths of the two beams are partially superimposed, then a macroscopic interference effect can be observed. Such beams are capable of interference and are called coherent.

Polarization of Light. Let us agree to represent an unpolarized beam of light by a straight-line arrow (Fig. 15 a), a polarized beam with vibrations in the plane of the paper by an arrow crossed by dashes (Fig. 15 b) and a polarized beam with vibrations perpendicular to the plane of the paper by an arrow with points (Fig. 15 c).

The following facts and regularities in connection with polarization of light have been established experimentally and theoretically (Fresnel theory).

1. Natural light (direct solar light and light from incandescent bodies) is unpolarized.

2. Natural light becomes partially polarized when it is reflected and refracted at a boundary between two transparent bodies in the case of oblique incidence (Fig. 16).

3. The relative amount of transmitted polarized light increases on repeated refraction. This fact has been used to construct piles of plates (Fig. 17) which may be used as polarizers to obtain almost fully polarized light, and as analyzers to detect polarization and also to determine the direction of the vibrations in polarized light. If a beam of light incident on such a pile at an oblique angle (best of all at Brewster's angle; see below) is unpolarized, then if the pile is rotated about the beam (at constant angle of incidence) the intensity of the transmitted light will remain constant. If partially polarized light is transmitted through the pile and the pile is rotated about the direction of the incident beam, the intensity of the transmitted light will be greatest when the direction of the vibrations in the beam coincides with the plane of incidence, and will have a minimum value when the two are perpendicular.

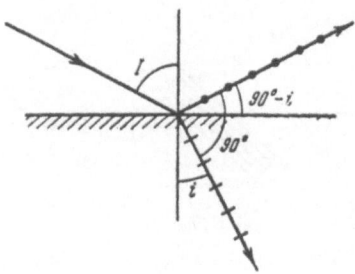

Fig. 18. Light reflected from the boundary between two transparent media will be totally polarized when the angle of incidence I is such that the reflected and refracted rays are at right angles.

4. In 1815 Brewster showed that reflected rays are completely polarized when the reflected and refracted rays are at 90° to each other. At the same time, the refracted rays have a maximum but not

total polarization (Fig. 18). The angle of incidence I in this case is known as Brewster's angle or the angle of total polarization.

It is important to note that in the case of total polarization of the reflected ray the dashes indicating the direction of vibrations in the transmitted ray are strictly parallel to the reflected ray. This partly shows that longitudinal vibrations in light propagated in an isotropic medium are impossible.

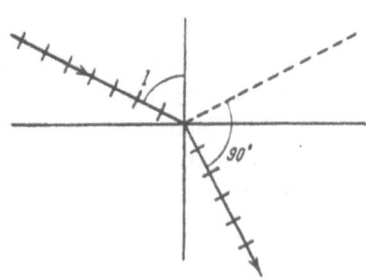

Fig. 19. Refraction of a polarized beam with vibrations in the plane of incidence.

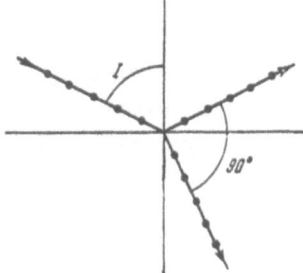

Fig. 20. Reflection and refraction of a polarized beam with vibrations perpendicular to the plane of incidence.

5. It is easy to show that the tangent of Brewster's angle is equal to the refractive index (Brewster's law), i.e.,

$$n = \operatorname{tg} I. \tag{35}$$

This follows from the fact that in this case $i = 90° - I$ and $\sin i = \cos I$.

6. Fully polarized light with vibrations in the plane of incidence, which strikes the boundary between two transparent dielectrics at Brewster's angle I, will be transmitted without partial reflection and will remain fully polarized in the same plane (Fig. 19).

7. Fully polarized light with vibrations in the plane perpendicular to the plane of incidence, which strikes the boundary at Brewster's angle I, is reflected and partly refracted but remains fully polarized in the same direction (Fig. 20).

DOUBLE REFRACTION

Double Refraction in Iceland Spar. Light is doubly refracted when it passes through noncubic crystals, liquid "crystals," and many other anisotropic bodies such as the fibres of plants and living tissues, hard glass, etc. This phenomenon was first studied in natural Iceland spar crystals (calcite, $CaCo_3$) by the Dutch scientist Erasmus Bartholinus in 1669.

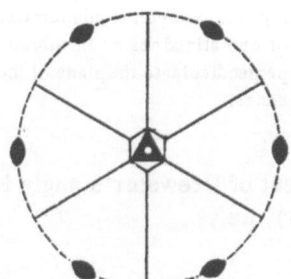

Fig. 21. The symmetry of an Iceland spar crystal. A sixfold mirror axis is perpendicular to the plane of the drawing. Symmetry planes are shown by straight lines. Three twofold axes lie in the plane of the drawing; each of them is shown by a pair of black "biangles." The center of symmetry is indicated by the white circle at the center of the drawing.

Iceland spar crystallizes in the $\bar{6} \cdot m$ class (Fig. 21). The crystals have natural cleavage planes along the $\{10\bar{1}1\}$ planes of a rhombohedron (Fig. 22). The calcite rhombohedron can be imagined as being produced from a cube by compressing the latter uniformly along one of its spatial diagonals. In the case of Iceland spar this diagonal is a fourfold axis (it is also a sixfold reflection and inversion axis) and is called the optic axis of the crystal.

The phenomenon of double refraction consists in the fact that all objects observed at short distance through a plane parallel plate of calcite, cut along the cleavage planes, appear double. If such a plate is placed on a piece of paper marked with an ink spot, two such spots will be seen through the crystal: one of them will remain stationary as the plate rotated in plane of the paper, and the other will describe a circle. If the plate is placed on a cardboard box with a hole illuminated from below, it is easily seen that the beam of light passing

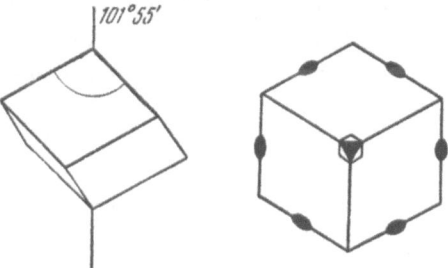

Fig. 22. A rhombohedron of Iceland spar (formed by six equal faces, each of which has the form of a rhombus).

through the aperture breaks into two beams within the plate, one of which consists of the so-called ordinary rays, and in the case of

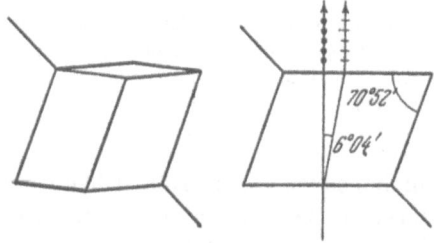

Fig. 23. Double refraction in Iceland spar. Light incident normally on the face of the rhombohedron. One of the symmetry planes of the rhombohedron lies in the plane of the drawing. The angle between the upper face and the right-hand edge is equal to 70° 52'. The ordinary ray passes through the rhombohedron without deviation along its entire path. The extraordinary ray is deviated through 6° 4' inside the crystal but leaves the crystal in a direction parallel to the direction of the ordinary ray. The rays leaving the crystal are polarized in mutually perpendicular planes.

normal incidence passes through the plate without refraction, and the other consists of extraordinary rays and is bent inside the crystal toward the optic axis through an angle of 6°4' but emerges from the plate parallel to the original beam (Fig. 23).

An examination of rays transmitted by the crystal, using a glass plate pile, shows that both the rays are fully polarized. The plane of vibrations of the ordinary rays is the plane containing the two rays (i.e., the plane of the diagram, which coincides with one of the symmetry planes of the crystal). The vibrations in the extraordinary ray are perpendicular to this plane (i.e., perpendicular to the plane of the diagram).

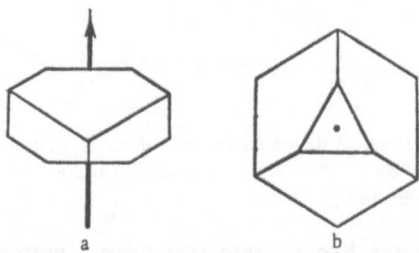

Fig. 24. A ray of light propagated in the direction of the principal axis of the rhombohedron and perpendicular to an artificially cut face perpendicular to this axis does not experience double refraction (a). This is also shown in the drawing on the right, in projection onto the artificially cut face (b).

If a calcite rhombohedron is cut so that it has two triangular {0001} planes perpendicular to the optic axis and the illuminated aperture in the cardboard box is observed through these planes, double refraction is not observed and the light transmitted in this direction is unpolarized (Fig. 24).

These observations lead us to the conclusion that calcite is optically anisotropic, i.e., it has different properties along different directions.

One detail of the phenomenon of double refraction described above is particularly important. It is well known that objects observed through a thick layer of water, e.g., at the bottom of an aquarium, appear to be nearer than they actually are. This is explained by saying that water has a larger refractive index than air. The two images of the aperture in the cardboard box observed through an Iceland spar crystal as described above also appear to be nearer. In addition, the image corresponding to the ordinary ray appears to be somewhat higher than the image corresponding to the extraordinary ray. This allows us to conclude that in calcite the refractive index for ordinary rays is greater than the refractive index for extraordinary rays, i.e.,

$$n_o > n_e.$$

Nicol prism. The difference in the refractive indices for ordinary and extraordinary rays in Iceland spar, and the ability of the latter

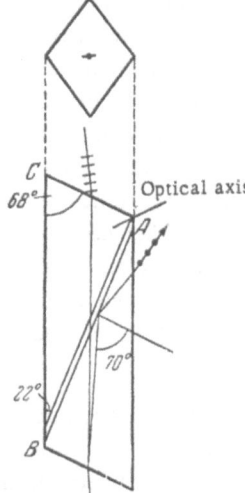

Fig. 25. The principle of a Nicol prism. A long prism is cut along a cleavage plane from a piece of Iceland spar and then divided into two parts along the plane AB. The two parts are then cemented together with Canada balsam. The ordinary ray which has a refractive index greater than that of the Canada balsam, is totally internally reflected from the balsam and leaves the prism through one side. The extraordinary ray, which has a lower refractive index, is transmitted by the prism. The vibrations in the extraordinary ray are parallel to the short diagonal of the rhombus.

to produce completely polarized light, was used by the Scottish physicist Nicol (1828) to produce polarizing prisms which are now called after him.

In order to produce a classical Nicol prism (Fig. 25) an Iceland spar rhombohedron is cut along a cleavage plane and then divided in two along the plane AB which is perpendicular to one of the symmetry planes of the rhombohedron which in Fig. 25 corresponds to the plane of the paper. The angle between the edge CB of the rhombohedron and the plane AB is made equal to 22°. The natural angle between CB and the natural surface AC, which is equal to 70°52', is then reduced to 68° and the two halves of the crystal are cemented together by means of Canada balsam.

The refractive index of Canada balsam is 1.54 and lies between the ordinary refractive index (1.658) and the extraordinary refractive index (1.516) in the direction along the length of the Nicol prism. The ordinary rays striking the Canada balsam layer are totally internally reflected (the angle of incidence exceeds the critical angle) because

Canada balsam has a lower refractive index, and emerge through the side of the prism and are absorbed by the prism holder. The extraordinary rays are transmitted by the prism without hindrance.

The classical Nicol prism has a small useful aperture (30°) and a considerable amount of material is required to make one of these prisms. When a Nicol prism is rotated about its longitudinal axis the beam of light transmitted by it becomes displaced since it does not emerge in the direction of the incident rays. Many other polarizing devices have been suggested in attempts to remove these disadvantages. The best of these is the Ahrens prism which has a large aperture and a small length in the direction of the incident light (Fig. 26).

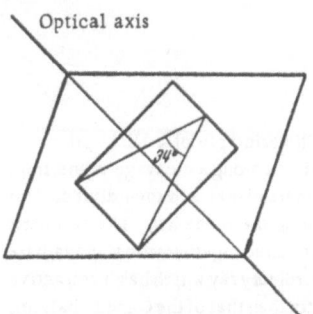

Fig. 26. The principle of an Ahrens prism. The optic axis and the symmetry plane of Iceland spar lie in the plane of the drawing. The Ahrens prism consists of three parts. The light is incident along the bisector of the 34° angle.

In recent years polaroid, invented by Land as far back as 1932, has been widely used and in many cases can replace crystal Nicol prisms. Polaroid consists of oriented submicroscopic crystals or complexes of long molecules of various organic substances. The action of these plates is based on unequal absorption of the two polarized waves which appear as a result of double refraction (pleochroism, p. 177). Light transmitted by polaroids (and also under certain conditions by Nicol prisms) is not completely polarized. White light transmitted by polaroid is slightly colored.

Rays and Wave Normals. As was explained above, when light passes from one isotropic medium to another its velocity and wavelength change. When it is incident normally on the boundary between the two media there is no refraction, and when it is incident obliquely, refraction is observed. The problem arises as to whether this rule, and the Descartes-Snell laws generally, are consistent with the fact that the extraordinary ray is bent away from the normal in the case of Iceland spar even for normal incidence. The problem is entirely resolved when one remembers that the laws of refraction in crystals refer essentially to wave normals or wave fronts rather than rays, and that in Iceland spar the wave surface is not spherical (p. 32). Thus the propagation of rays of light inside a calcite crystal must be as shown in Fig. 27 a, or in a more general case (for crystals of lower systems)

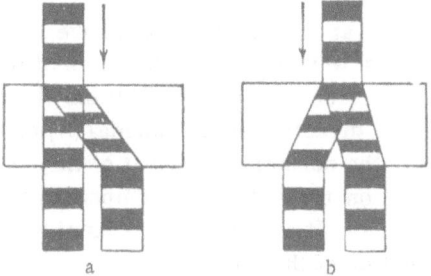

a   b

Fig. 27   Double refraction in uniaxial (a) and biaxial (b) crystals. When light is incident normally on the crystal plate, only the rays are refracted
but not the waves (wavefronts) nor the wave normals.

as in Fig. 27 b. These figures show that refraction is absent, i.e., the waves outside and inside the crystal are parallel. This is also true of the normals to the wave fronts which under these conditions are not refracted (normal incidence) as was to be expected on the basis of the Descartes-Snell laws.

We have thus come to a very important conclusion: the concepts of wave normals and rays, which in the case of isotropic media are identical, refer to very different things in the case of optically anisotropic media. It is clear from the above drawings that a ray represents the direction along which the light vibrations and hence the luminous energy are actually transmitted. The eye will only see a source of light if it looks along the direction of a ray and not a wave normal. The physical meaning of the wave normal will now be explained.

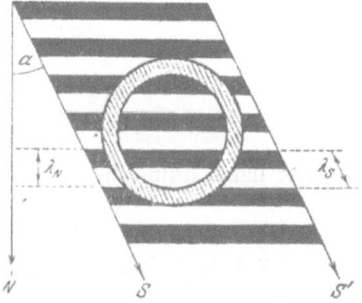

Fig. 28. The concept of a ray and a wave normal. When the true motion of the waves is in the direction of the ray S, an observer looking at a limited part of the wave (through the circular window) will see a wave motion in the direction of the wave normal N. The wavelength in the direction of the normal (λ_N) is smaller than the wavelength in the direction of the ray (λ_S). If the angle between N and S is equal to α, then $\lambda_N = \lambda_S \cos \alpha$.

29

Consider a beam of plane waves which is represented by a system of black and white bands in Fig. 28 and which is bounded by the two parallel rays S, S' and moves in their direction. We shall imagine an observer who looks at this picture from above through a wide aperture placed in such a way that the boundaries of the beam cannot be seen. It is clear that under these conditions the observer will see waves moving in the direction of the normal N, or more accurately, he will record only the normal component of the true motion and will have no information on the tangential component (he will not even be aware of its existence). It is this wave normal which is taken as the direction of propagation of the waves.

Quantities Used to Describe Wave Motion in Anisotropic Media. It is clear from the above discussion that in the case of anisotropic media we must distinguish between the wavelength λ_S measured along the ray direction and the wavelength λ_N measured along the wave normal, and also between the ray velocity v_S and the normal velocity v_N. It is clear from Fig. 28 that the relation between λ_N and λ_S is

$$\lambda_N = \lambda_S \cos \alpha, \tag{36}$$

and the relation between v_N and v_S is

$$v_N = v_S \cos \alpha. \tag{37}$$

Hence

$$\frac{v_N}{v_S} = \frac{\lambda_N}{\lambda_S}. \tag{38}$$

On the other hand, using (6) we have

$$v_N = \nu_N \lambda_N,$$
$$v_S = \nu_S \lambda_S, \tag{39}$$

where ν_N, ν_S are the frequencies corresponding to wave normal and ray, respectively. It follows that

$$\nu_N = \nu_S,$$

i.e., the frequencies in the ray and the associated wave normal are equal. We shall therefore denote the frequency by the symbol ν without subscripts

$$\nu_N = \nu_S = \nu. \tag{40}$$

We saw earlier (12') that in the case of isotropic media

$$n = \frac{c}{v} = \frac{\lambda_0}{\lambda}.$$

Since in the case of anisotropic media the Descartes-Snell laws are obeyed by the wave normals and not the rays, the above equation will in this case take the form

$$n = n_N = \frac{c}{v_N} = \frac{\lambda_0}{\lambda_N}.\tag{41}$$

We thus see that by the refractive index one means the quantity n_N and not the quantity

$$n_S = \frac{c}{v_S} = \frac{\lambda_0}{\lambda_S}.\tag{42}$$

The meaning of n_S will be explained below (p. 50).

Direction of Vibrations in the Extraordinary Ray. Having discussed the concepts of rays, velocities, wavelengths and refractive indices in the case of anisotropic media we must, clearly, also consider the concept of the direction of vibrations. It was pointed out ear-

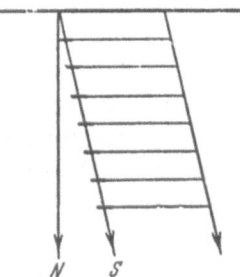

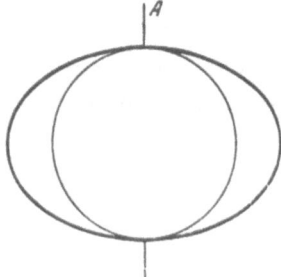

Fig. 29. The direction of vibrations in the ray S lies in the plane N_S and is perpendicular to N.

Fig. 30. A wave surface of two sheets in Iceland spar, according to Huygens. A) An infinite-fold axis.

lier that for plane waves propagated in an isotropic medium the light vector vibrates perpendicularly to the ray and hence parallel to the wave front. All the phenomena considered in optical crystallography can be very well explained on the assumption that in anisotropic media the light vibrations always lie in the wave front. This means that the vibrations in the extraordinary rays are perpendicular to both the ray and the wave normal, while in the case of extraordinary rays they lie in the plane formed by the ray and the corresponding wave normal and are perpendicular to the wave normal but not to the ray (Fig. 29).

The Wave Surface in Iceland Spar According to Huygens. If we choose any point within a crystal of Iceland spar, and imagine lines radiating from this point in all directions such that their lengths are equal or proportional to the ray velocities v_S, then the ends of these

31

lines will describe a surface which is known as the wave surface or ray velocity surface. Since in a doubly refracting crystal there are two rays travelling with different velocities along each direction, it follows that this surface must, in fact, be a surface of two sheets. Clearly, the wave surface may be defined as the surface reached in one second by light emitted from a point source located inside the crystal.

Having studied the optical properties of calcite, the Dutch physicist Huygens, who was a supporter of the wave theory of light, came to the conclusion that the above surface has the form of an elongated ellipsoid of revolution with an inscribed sphere (Fig. 30). The line connecting the points of contact between the sphere and the ellipsoid is an infinite symmetry axis and also the optic axis of the crystal. Since the ray velocities v_S are different for different colors, the corresponding ray surfaces are also somewhat different. However, the optic axes of all these surfaces coincide.

Huygens Constructions. If one knows the form of the wave surface and its position within the crystal it is easy to construct the rays and the wave normals for any given angle of incidence. This construction was first put forward by Huygens (who did not consider the problem of the direction of vibrations since at that time light vibrations were thought to be longitudinal). We shall give a few examples of the Huygens construction.

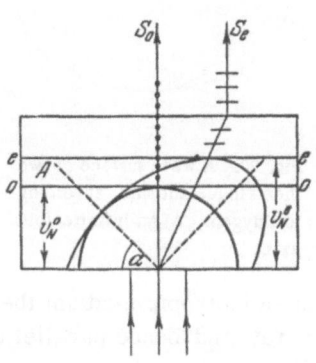

Fig. 31. Huygens construction for Iceland spar. The rays are incident normally on a plane cut at an angle α to the optic axis A of the crystal.

Consider a monochromatic unpolarized beam of light incident normally (from air) on a calcite plate cut at an angle α to the optic axis of the crystal A. Having drawn the corresponding profile of the wave surface for one of the incident rays (Fig. 31) we can find the directions S_o and S_e of the ordinary and extraordinary rays both inside and outside the plate, and also the direction of vibrations in each of them. The construction is based on the fact that the waves entering and leaving the plate are not refracted, and the fact that the ordinary ray moving with the velocity $v_N^o = v_S^o$, will reach the tangent

$o - o$ to the circle in the same time in which the extraordinary wave whose normal velocity is v_N^e, will reach the tangent $e - e$ to the ellipse. The directions of vibrations in both waves should, as was pointed out above, be parallel to the wave surfaces and lie in the plane defined by the normal and the ray.

So as not to complicate the drawing we considered only one of the rays in the incident beam and showed that the rays emerging from the plate (S_o, S_e) travel in the same direction but along different paths. It is easy to show, however, that if the construction were completed for all the rays in the incident beam, then provided the width of this beam is sufficiently large, the emergent ordinary and extraordinary rays would be partly superimposed. In the common region, the path of each extraordinary ray will coincide with the path of an ordinary ray.

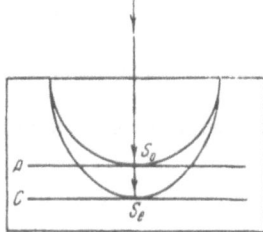

Fig. 32. Huygens construction for Iceland spar. The rays are incident normally on the face of the prism and travel within the crystal in a direction perpendicular to the optic axis.

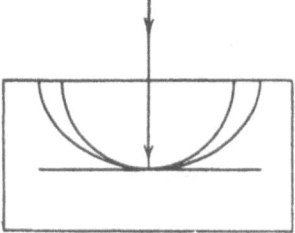

Fig. 33. Huygens construction for Iceland spar. The rays are incident normally on the face of the crystal and travel within it along the optic axis. Double refraction is absent.

Let us suppose now that the plate of Iceland spar is cut parallel to the optic axis (Fig. 32) and an unpolarized beam is incident normally on it from air. It is easy to see from the Huygens construction that in this case the ray will not be refracted. However, its separation into two differently polarized rays S_o and S_e, moving along the same path but with different velocities will still take place.

As was pointed out above, it has been established experimentally that an unpolarized beam incident normally onto a calcite plate cut perpendicularly to the optic axis does not experience any real changes and remains unpolarized and is not divided into two rays. A reference to the Huygens construction shown in Fig. 33 will make this clear.

We have considered three simple special cases of the Huygens construction. Let us now consider a more complicated and more general case.

Consider a plane wave of natural light AB incident on a calcite crystal from air at an angle i_1, and suppose that the optic axis of the crystal lies in the plane of the diagram (Fig. 34). Let us further assume

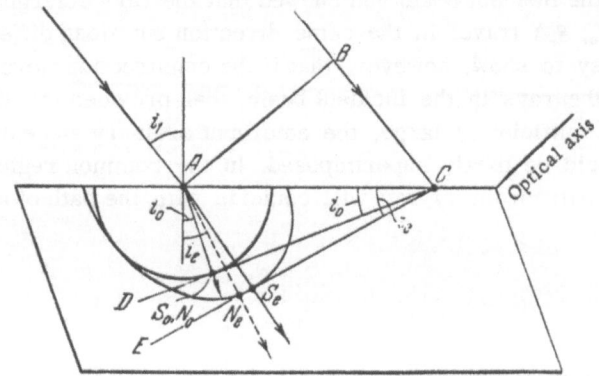

Fig. 34. Huygens construction for Iceland spar. Plane of incidence coincides with the plane of symmetry of the rhombohedron of Iceland spar and the plane of the drawing. The rays are incident obliquely.

that the path BC is traversed in unit time, i.e., $BC = c$. We then draw the wave surfaces with a center at A and in accordance with the given orientation of the optic axis, and then construct the tangents DC and EC to the circle and the ellipse. We then connect the point A to the point of contact S_0 (which is also the N_0) of the line DC and the circle and thus find the direction of the ordinary ray. We then connect the point A to the point of contact S_e' of the line EC and the circle, thereby finding the direction of the extraordinary ray; the corresponding wave normal is perpendicular to the straight line EC drawn from the point A. We then have

$$BC = c,$$
$$AS_0 = AN_0' = v_N^0 = v_S^0,$$
$$AS_e = v_S^e,$$
$$AN_e = v_N^e.$$

From the triangles ABC, ACN_0, and ACN_e we have

$$c = AC \sin i_1,$$
$$v_N^0 = AC \sin i_0,$$
$$v_N^e = AC \sin i_e,$$

and hence

$$\frac{c}{v_N^0} = \frac{\sin i_1}{\sin i_0} = n_o,$$

$$\frac{c}{v_N^e} = \frac{\sin i_1}{\sin i_e} = n_e.$$

The above construction confirms the earlier assumption that the Descartes-Snell laws, which hold for optically isotropic bodies, can be generalized to optically anisotropic bodies if the ray velocities are replaced by normal velocities.

Fresnel's Theory. Fresnel (1821 – 1827) considered that the reason for the phenomenon of double refraction in crystals is that the elasticity of the ether in crystals is different along different directions. He thought that light waves are transverse and suggested that the ray

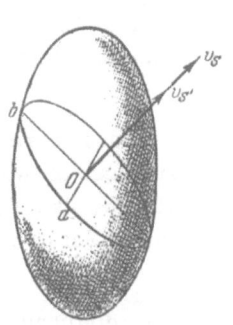 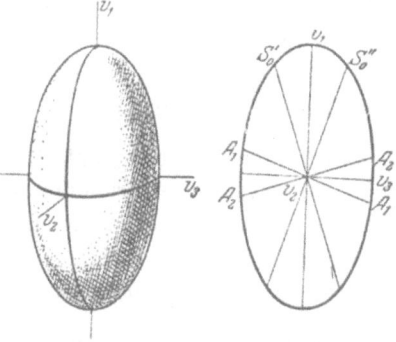

Fig. 35. Principle of construction of the ray velocity surface of two sheets (v_S). The Fresnel ellipsoid is cut by a plane passing through the center of the ellipsoid. This section has the form of an ellipse with principal semi-axes a, b which are numerically equal to $v_{S'}$, $v_{S''}$. If one draws vectors $v_{S'}$ and $v_{S''}$ perpendicular to the plane of the ellipse, one finds the ray velocity in this direction, and the end points of the vectors define two points on the required surface. All the other points on the surface are found in a similar way.

Fig. 36. General Fresnel ellipsoid. Principal semiaxes of the ellipsoid are numerically equal to the principal velocities. The true velocities v_1, v_2 are in the direction of the v_3 axis, the velocities v_1, v_3 are along the v_2 axis and the velocities v_2, v_3 along the v_1 axis. Two circular sections A_1A_1 and A_2A_2 pass through the mean axis of the Fresnel ellipsoid. The biradials S'_o, S''_o are perpendicular to them.

velocity v_S in a given direction should be proportional to the square root of the elasticity of the ether in the transverse direction, or more accurately, in the direction in which the vibrations take place. Ac-

35

cording to Fresnel, the ray velocity v_S is given by the formula

$$v \approx \sqrt{\frac{N}{\rho}},$$

(which also represents the velocity v of the propagation of transverse elastic waves in a solid medium) provided the velocity v is replaced by the ray velocity v_S, the rigidity N is replaced by the elasticity of the ether Q and the density ρ by the density of the ether r (p. 53). Fresnel assumed that, similarly to the other properties of crystals, the quantity $\sqrt{Q}$ varies with direction according to an ellipsoidal law. Since for each radius vector of the ellipsoid there are two special transverse directions, namely, the two principle axes of the corresponding elliptical cross section of the ellipsoid, it follows that there are two ray velocities v_S' and v_S'' (these were earlier denoted by v_S^o and v_S^e) which are proportional to the semiaxes oa and ob of the ellipse (Fig. 35). The wave surface of two sheets may be obtained by plotting the values of v_S' and v_S'' along different directions starting from a fixed point in the crystal.

The Fresnel theory explained the existence of double refraction in calcite crystals and indicated how the optical surfaces of two sheets of other crystals could be constructed theoretically. We shall describe these constructions in a later section; to begin with let us consider the properties of the above Fresnel ellipsoid.

Properties of the Fresnel ellipsoid. We must distinguish three types of Fresnel ellipsoids, namely, the general ellipsoid, the ellipsoid of revolution and the sphere. The general ellipsoid has the following symmetry elements: three twofold axes, three symmetry planes and a center of symmetry (Fig. 36), i.e., it belongs to the $m \cdot 2 : m$. symmetry group. The principle axes of the ellipsoid whose lengths are v_1, v_2, v_3 coincide with the symmetry axes and are equal to the principal velocities but not to each other[*]. It is essential to remember that in accordance with the Fresnel theory, light is propagated with with two velocities (v_2 and v_3) along the v_1 axis. Similar considerations apply in the case of the other axes. Next, it is important to remember that from symmetry considerations there is no difference between the velocities v_i^S and v_i^N in directions corresponding to the principal axes. For this reason, the principal velocities are given only the numerical indices 1,2,3 and the indices N and S are omitted.

---

[*] By the velocity of light we shall understand, both here and in the subsequent paragraphs, the phase velocity of monochromatic light.

The general ellipsoid has two circular cross sections, A_1A_1 and A_2A_2, which intersect along the v_2 axis. We may verify that they exist in the following way. Any arbitrary cross section of an ellipsoid is, in general, an ellipse. Consider one of the elliptical cross sections which includes the v_2, axis, and vary the angle between this plane and the major axis v_3 or the minor axis v_1. In this way the v_2 axis of the ellipse will remain unaltered while the other axis will lie between v_1 and v_3. It is clear that at some intermediate position of the ellipse its second axis will also be equal to v_2. This cross section will therefore be a circular one.

The normals S_0' and S_0'' to the circular cross sections are known as biradials or ray optic axes.

When $v_3 > v_2 = v_1$ we have a prolate ellipsoid of revolution and when $v_3 < v_2 = v_1$ the ellipsoid is an oblate ellipsoid of revolution. Each of them has only one circular cross section and hence only one optic axis. We shall see later that there is no need to call it a ray optic axis. It is an infinite-fold axis of symmetry of the ellipsoid. Each ellipsoid of revolution has an infinite number of longitudinal symmetry planes, one transverse symmetry plane, an infinite number of transverse fourfold axes of symmetry and a center of symmetry. The corresponding symmetry group is denoted by the symbol $m \cdot \infty : m$.

When $v_1 = v_2 = v_3$ the ellipsoid reduces to a sphere having a $\infty/\infty \cdot m$ symmetry.

Classification of Crystals According to the Form of the Fresnel Ellipsoid. Crystals whose optical properties can be described by a spherical Fresnel surface are optically isotropic. Crystals whose optical properties can be described by an ellipsoid of revolution are optically anisotropic and are called optically uniaxial. All other crystals are optically biaxial and also optically anisotropic, and their optical properties are described by the general ellipsoid.

It is known from elementary crystallography that a given crystal may have different symmetries for different properties. If the particular symmetry is not specified then it is understood that one deals with the lowest symmetry of the crystal which we shall call the morphological symmetry since, in principle, it can always be established by studying the morphological properties of the crystal. The morphological symmetry of a crystal should, of course, never be higher than the optical symmetry, and should be subordinate to the nearest optical symmetry, i.e., it should either coincide with the lowest of the three possible symmetry groups of the Fresnel ellipsoid or should be its subgroup.

This means that cubic crystals have no special directions* and are optically isotropic. Their symmetry is subordinate to the spherical group.

Crystals belonging to the hexagonal, tetragonal and trigonal systems are optically uniaxial. They have only one special direction, namely, the optic axis which coincides with the principal axis of symmetry of the crystal. Their symmetry is subordinate to the symmetry of an ellipsoid of revolution. Uniaxial crystals whose optical properties are described by an oblate Fresnel ellipsoid are known as optically positive. Optically negative crystals correspond to prolate Fresnel ellipsoids. We note, however, that a given crystal may be positive for one frequency and negative for another, and moreover, it can change its sign, e.g., on heating. Therefore there is no special physical basis for dividing crystals into two categories according to their optical sign.

Crystals belonging to the lower systems (orthorhombic, monoclinic and triclinic) are optically biaxial and have two optic axes (pp. 43 and 44). Each of them has more than one special direction. All the symmetry groups are subgroups of the $m \cdot 2 : m$; group, except for one which coincides with it. Each symmetry element of these groups should, accordingly, coincide with the corresponding symmetry element of the general ellipsoid. Biaxial crystals can also differ in their optical sign (p. 45).

There are homogenous media such as liquid crystals, glass in a uniform electric or magnetic field, polaroids, etc., which are not monocrystals but have the optical properties of the latter. The symmetry of such media at each given point can be determined through a comparison of its various properties. In principle this symmetry can be arbitrary. In particular, we can imagine, and even make, artificially homogenous media which have fivefold, sevenfold and higher symmetry axes, which are impossible in the case of crystals. In fact, there are natural media with infinite-fold symmetry axes. All such media can be optically isotropic, uniaxial or biaxial just as crystals.

---

*In this case by "direction" we simply mean a straight line drawn through a special point within the crystal. If there are many special points within the crystal any one of them may be taken. Special points and directions within a crystal are defined to be those which cannot be repeated by symmetry operations.

Construction of the v_S Surface from the Fresnel Ellipsoid.

By the v_S surface we shall understand a surface whose radius vectors are equal (or proportional) to the ray velocities in the crystal. As was pointed out above, this surface is reached in a given interval of time by light emitted by a point source located inside the crystal. The form of this surface corresponds to the real form of the light wave and hence the v_S surface is known as the wave surface. The principle of constructing the wave surface for different crystals using the Fresnel ellipsoid was given above and we shall now describe the construction in more detail.

1. It is known a priori that the wave surface of optically isotropic crystals cannot differ from the wave surface of other optically isotropic bodies and should be a sphere.

2. In order to construct the wave surface for uniaxial positive crystals we must start (as we already know) from an oblate ellipsoid

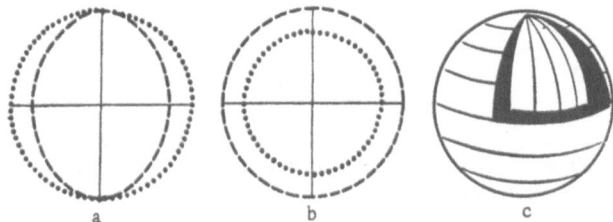

a   b   c

Fig. 37. Wave surface for uniaxial positive crystals: a sphere with an inscribed prolate ellipsoid of revolution. a) Principal section (the optic axis lies in this section); b) section perpendicular to the optic axis; c) three-dimensional representation. The vibration directions are indicated by dashes and dots; in the three-dimensional representation they are represented by meridians and parallels.

of revolution, and set out along the perpendicular to each central cross section of the ellipsoid, quantities proportional to the principal axes of the cross-sectional ellipse. For all these cross sections, the semimajor axis, which is also the radius of the circular cross section of the ellipsoid, will remain constant and equal to $v_1 = v_2$. The other semiaxis will lie between v_1 and v_3 depending on the orientation of the cross section. Accordingly, one part of the v_S surface will be a sphere having a radius v_1, and the other part will be a prolate ellipsoid of revolution inscribed in the sphere (Fig. 37). The axis of revolution of this ellipsoid coincides with the axis of revolution of the original Fresnel ellipsoid. The spherical surface corresponds to the ordinary rays and the ellipsoidal surface to extraordinary rays. Light vibrations in the extraordinary rays are in the meridional direction and the vibrations in the ordinary rays in the equatorial direction. This is a consequence of the Fresnel theory according to which the direction of vibrations in a ray S is obtained by constructing at the end of the v_S vector a plane tangential to the wave surface, and dropping a perpendicular from the center of the surface to this plane, and then drawing a straight line through the end of the perpendicular and the end point of the vector v_S. The direction of this straight line is the direction of the vibrations in the ray S. In the drawing, the direction of vibrations in the plane sections through the ellipsoid is shown by dots and dashes, while in the three-dimensional representation, the direction of vibrations is indicated by the meridians and parallels.

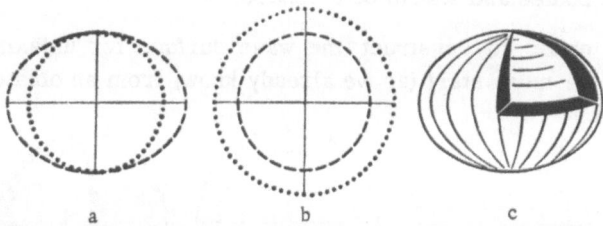

a   b   c

Fig. 38. Wave surface of uniaxial negative crystals–oblate ellipsoids of revolution with an inscribed sphere. Vibration directions are indicated as in Fig. 37.

In the case of uniaxial negative crystals, the Fresnel construction leads to the wave surface which we have already encountered in the case of calcite and which consists of an oblate ellipsoid of revolution and an inscribed sphere (Fig. 30). Here this surface is shown in two sections and a three-dimensional drawing (Fig. 38). Both in the case of negative and positive surfaces, the sphere corresponds to ordinary and the ellipsoid to extraordinary rays. The vibrations in the ordinary rays are in the equatorial direction, and in the extraordinary rays they are in the meridional direction.

Quartz is an example of a uniaxial positive crystal and Iceland spar is an example of a negative uniaxial crystal.

3. To construct the v_S surface for a biaxial crystal we start with a general ellipsoid. Consider the three principal sections through this surface (Fig. 39). In order to construct the section perpendicular to the v_2 axis of the Fresnel ellipsoid we proceed as follows. First we cut the ellipsoid along the v_2v_3 plane. We then set out segments equal in length to its semi-axes v_2, v_3, on either side along the v_1 axis of the

Fresnel ellipsoid, i.e., along the X_1 axis of the required surface. By rotating the cutting plane about the v_2, axis we obtain new elliptical

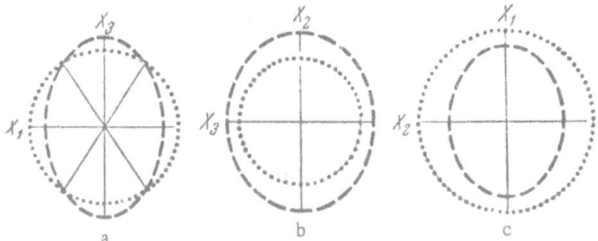

a   b   c

Fig. 39. Three sections through the wave surface of biaxial crystals. a) Perpendicular to the mean axis of the Fresnel ellipsoid; b) perpendicular to the minor axis of the Fresnel ellipsoid; c) perpendicular to the major axis of the Fresnel ellipsoid.

sections of the ellipsoid with a constant principal semiaxis v_2 and a variable axis v_S, which lies between the semiminor axis v_1, and the semimajor axis v_3 of the ellipsoid. If we set out along the two normals to the plane of each of these ellipses, segments equal to their principal semiaxes, we obtain a circle with radius v_2, intersecting the ellipse with principal semiaxes v_1 and v_3 (Fig. 39 a). The other two sections of the required v_S surface are obtained in a similar way. The section perpendicular to the semiminor axis v_1 of the Fresnel ellipsoid is an ellipse with semiaxes v_2, v_S and a circle having a radius v_1 and located inside the ellipse as shown in Fig. 39 b. The section perpendicular to the semimajor axis v_3 consists of a circle having a radius equal to the semimajor axis v_3 of the Fresnel ellipsoid, and an ellipse with semiaxes equal to v_1 and v_2 and lying inside the circle (Fig. 39 c). The directions of vibrations at all points of these sections are shown as usual by dots and dashes.

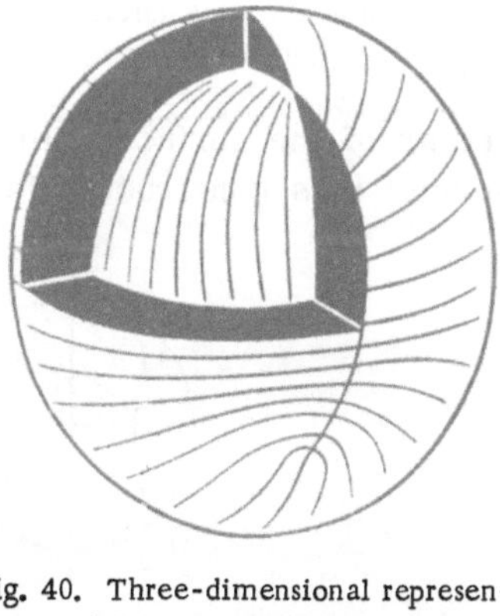

Fig. 40. Three-dimensional representation of the wave surface of a biaxial crystal. Vibration directions of the inner surface have a meridional character and the vibration directions of the outer surface have an equatorial character.

The general form of the wave surface in biaxial crystals is given in Fig. 40. As can be seen, this is a complex surface formed by two component surfaces which intersect at four points lying in funnel-shaped indentations. The biradials pass

41

through these points and the center of the surface. There is no double refraction in the direction of the biradials. Since the two parts of the surface are nonspherical, the rays which appear in biaxial crystals as a result of double refraction are both extraordinary rays.

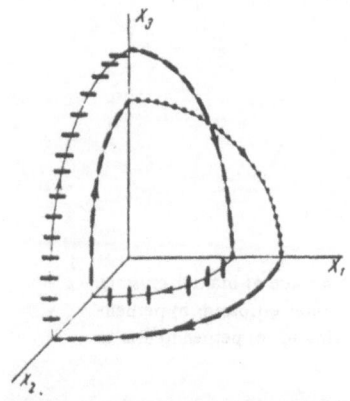

It is interesting to note that an octant of the above surface may be drawn by a continuous motion of the pen, since elliptical arcs in this drawing (Fig. 41) alternate with circular arcs. On the elliptical arcs the vibrations are along the tangents and on the circular arcs the vibrations are perpendicular to the planes containing the arcs.

Fig. 41. One of the octants of the wave surface of a biaxial crystal drawn by a single movement of a pen along the directions indicated by the arrows. Circular arcs alternate with elliptical arcs. Motion along the ellipses is accompanied by a change in the vibration directions while motion along the circle is not accompanied by such changes.

Construction of the v_N Surface. We already know that in order to construct a wave normal corresponding to a given ray we must construct a plane tangential to the wave surface at the point where the ray intersects this surface, and drop a perpendicular from the center of the surface onto the tangential plane. The direction of the normal thus constructed is the same as the direction of the normal

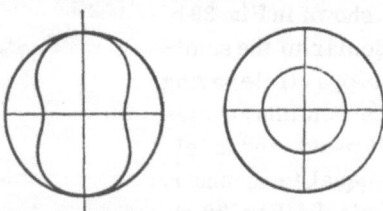

Fig. 42. Normal velocity surface of a positive crystal. Sphere with an inscribed prolate ovaloid. Two sections through the surface are shown.

velocity v_N and its length is equal to the magnitude of this velocity, while the end point of the normal lies on the v_N surface, or more accurately, the normal velocity surface. All the other points of this surface are found in a similar way.

In the case of optically isotropic crystals, the v_N and v_S surfaces are identical and are both spheres.

The v_N surface for uniaxial positive crystals consists of a sphere with an inscribed prolate ovaloid of revolution (Fig. 42). The figure gives two sections through this surface, namely, parallel and perpendicular to the optic axis. The first of these consists of a circle with an inscribed oval and the second consists of two concentric circles. The method of constructing the oval from an ellipse is shown in Fig. 43.

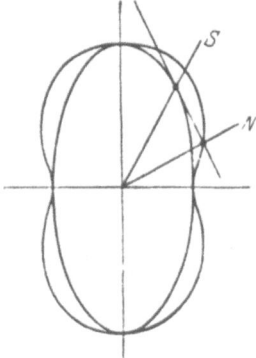

The v_N surface for uniaxial negative crystals consists of an oblate ovaloid of revolution with an inscribed sphere. Figure 44 shows two sections through this surface.

The two v_N surfaces for uniaxial crystals show the main characteristics of the corresponding v_S surfaces. The only difference is that in the case of the v_N surfaces the ellipsoids are replaced by ovaloids.

Fig. 43. The construction of a normal velocity surface from a wave surface. The drawing shows the construction of an oval from an ellipse.

This statement also holds in the case of the v_N surfaces of biaxial crystals (Fig. 45 a). The section of this surface which is perpendicular

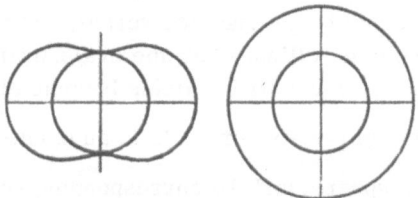

Fig. 44. Normal velocity surface of a uniaxial negative crystal. Oblate ovaloid of revolution with an inscribed sphere. Two sections through the surface are shown.

to the mean axis v_2 of the Fresnel ellipsoid deserves special attention (Fig. 45 b). It consists of an oval intersecting a circle at four points. The straight lines N' and N'' which connect these points to the center of the oval are called the binormals, or simply optic axes of the biaxial crystal (without indicating that these axes have the direction of the normals and not the rays).

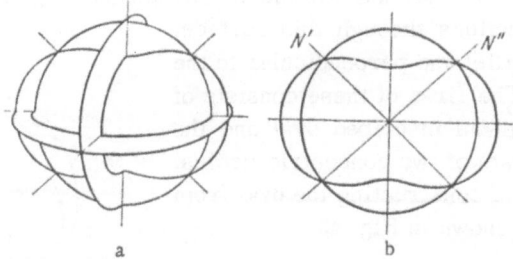

Fig. 45. Normal velocity surface of a biaxial crystal. a) In an oblique position; b) section containing the optic axes N', N''.

Construction of the $\frac{1}{v_N}$ *and* $\frac{1}{v_S}$ *Surfaces.* The $\frac{1}{v_N}$ surface is the locus of the end points of radius vectors whose lengths are proportional to the refractive indices and is sometimes also called the index surface. It follows from its definition that the surface may be constructed from a v_N surface by replacing all the radius vectors of the v_N surface by their reciprocals, or what amounts to the same thing, by quantities proportional to the refractive indices. The $\frac{1}{v_S}$ surface can be constructed in a similar way.

When the radius vectors are replaced by their reciprocals, small radius vectors become large (and vice versa), and prolate figures are transformed into oblate, ellipses become ovals, ovals become ellipses, ellipsoids become ovaloids, and ovaloids become ellipsoids. Thus the $\frac{1}{v_N}$ surface of uniaxial positive crystals is an oblate ellipsoid of revolution containing a sphere, and the corresponding surface for negative uniaxial crystals is a prolate ellipsoid of revolution inscribed in a sphere. In general, the $\frac{1}{v_N}$ surfaces will be similar in form to the v_S surfaces of crystals of the opposite sign.

The Optical Indicatrix. We used the Fresnel ellipsoid to construct the surfaces of two sheets. It is, however, easy to show that instead we could have used another ellipsoid, known as the optical indicatrix, whose semiaxes are proportional to the principal refractive indices n_3, n_2, n_1;[*] which in turn are proportional to $\frac{1}{v_3}, \frac{1}{v_2}, \frac{1}{v_1}$, as we already know.

The optical indicatrix is an ellipsoid and has, in general, two special circular sections passing through its center. The normals to these circular sections are the binormals or optic axes which we have already met.

The angle $2V$ between the binormals is called the optic axial angle. The half-angle V can be calculated from the formula

$$\operatorname{tg} V = \sqrt{\frac{v_3^2 - v_2^2}{v_2^2 - v_1^2}} = \sqrt{\frac{\dfrac{1}{n_3^2} - \dfrac{1}{n_2^2}}{\dfrac{1}{n_2^2} - \dfrac{1}{n_1^2}}}. \tag{43}$$

and is measured from the major axis n_1 of the indicatrix. This formula may be derived using Fig. 46 by solving simultaneously the equation of the ellipse

$$\frac{x_1^2}{n_1^2} + \frac{x_3^2}{n_3^2} = 1$$

and the equation of the straight line

$$\operatorname{tg} V = \frac{x_3}{x_1}.$$

It was pointed out above that biaxial crystals can differ in optical sign. If $V<45°$, the crystal is positive and when $V>45°$ it is negative. In other words, in the case of positive crystals the major axis n_1 of the indicatrix is the acute bisectrix, i.e., it bisects the acute angle between the optic axes, and

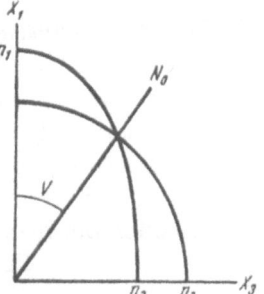

Fig. 46. Concerning the derivation of Formula (43).

in negative crystals it is the minor axis n_3 which does so. The relations which hold for uniaxial crystals are opposite to those which were established in connection with the Fresnel ellipsoid, i.e., positive crystals are characterized by a prolate indicatrix of revolution and negative crystals by an oblate one.

---

[*] In petrographic practice, the minimum, intermediate and maximum refractive indices are denoted, following Fedorov, by the symbols N_p, N_m, N_g. The letters p, m, g are the first letters of the French words petit (small), moyen (intermediate) and grand (large).

If the Fresnel construction is carried out using the indicatrix directly, the $\frac{1}{v_N}$ surface can be immediately obtained. Other surfaces of two sheets can be obtained by simple constructions which we need not consider here.

The Velocity Ovaloid and the Index Ovaloid. The construction of surfaces of two sheets can also be based on a simple surface known as the velocity ovaloid whose principal semiaxes are v_1, v_2, and v_3. This ovaloid can be obtained from the Fresnel ellipsoid by dropping perpendiculars from the center of the ellipsoid to the tangential planes at all points on the ellipsoid, and the required ovaloid is the locus of the points of intersection of these perpendiculars with the tangential planes.

Similarly, the optical indicatrix may be used to obtain the index ovaloid, i.e., an ovaloid having n_3, n_2, and n_1, as its principal axes. Using the Fresnel construction, the $\frac{1}{v_S}$ surface of two sheets may be obtained directly from this ovaloid.

The Connection between All the Above Surfaces. Although each of the above surfaces can be used to construct all the others, the simplest and most natural connection between them, allowing the most convenient transition from one to another, is shown in the following table.

Connection Between the Eight Surfaces Describing Double
Refraction in Crystals

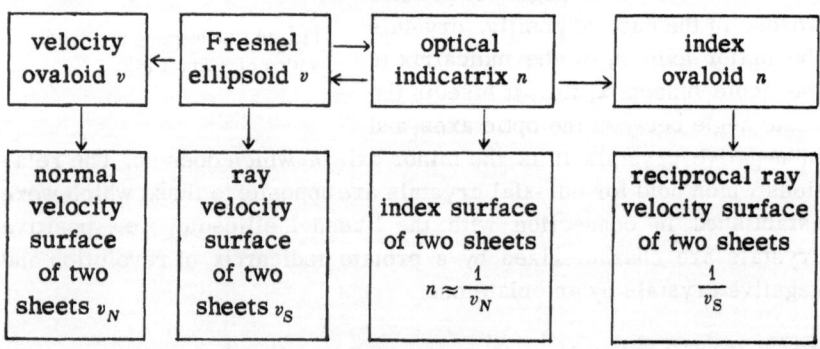

Although, so far, only three of the above eight surfaces have been widely used, namely, the Fresnel ellipsoid v, the optical indicatrix n, and the wave surface v_S, double refraction in crystals can only be well understood by using all the eight surfaces and their interdependence.

Equations of Optical Surfaces. The Fresnel ellipsoid has semiaxes equal to v_1, v_2, and v_3 and hence its equation in a Cartesian system of coordinates is

$$\frac{x_1^2}{v_1^2} + \frac{x_2^2}{v_2^2} + \frac{x_3^2}{v_3^2} = 1. \tag{44}$$

In physical crystallography, special polar coordinates are often used to describe the various surfaces. These coordinates are the four variables r, α_1, α_2, and α_3. The variable r is the radius vector of the surface drawn from the origin to a point on the surface, and the variables α_1, α_2, and α_3 are the angles between the radius vector and the X_1, X_2, and X_3 axes. These angles are given by the direction cosines c_1, c_2, and c_3, i.e., the cosines of angles between the radius vector and the axes of a rectangular set of coordinates.

Since the direction cosines are related by the expression

$$c_1^2 + c_2^2 + c_3^2 = 1, \tag{45}$$

only three of the above four variables are independent. The above set of coordinates is therefore three-dimensional, although at first sight it appears to be four-dimensional.

It is clear from Fig. 47 that the relation between the Cartesian coordinates x_1, x_2, and x_3 of the end point of the radius vector r and the polar coordinates is given by the equations

$$r = \sqrt{x_1^2 + x_2^2 + x_3^2},$$

$$x_1 = rc_1; \quad x_2 = rc_2; \quad x_3 = rc_3,$$

which may be used to transform easily from one system to the other.

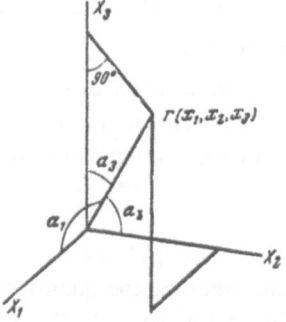

Using these relations, and bearing in mind the fact that in the Fresnel ellipsoid $r = v_S$, the equation of this ellipsoid is

Fig. 47. Concerning the derivation of the equations of optical surfaces.

$$\frac{1}{v_S^2} = \frac{c_1^2}{v_1^2} + \frac{c_2^2}{v_2^2} + \frac{c_3^2}{v_3^2} \tag{46}$$

The equation of the optical indicatrix can be obtained from (45) and

47

(46) by replacing the velocities v_1, v_2, v_3 by the refractive indices n_1, n_2, n_3.

The equation for the ovaloid with semiaxes v_1, v_2, v_3 and radius vectors r equal to v_N has the form

$$v_N^2 = c_1^2 v_1^2 + c_2^2 v_2^2 + c_3^2 v_3^2.$$ (47)

The latter equation can be derived in the following way. The equation of the plane tangential to the ellipsoid at the point whose coordinates are a_1, a_2, a_3 is:

$$\frac{x_1 a_1}{v_1^2} + \frac{x_2 a_2}{v_2^2} + \frac{x_3 a_3}{v_3^2} = 1;$$ (48)

and the equation of the perpendicular dropped from the center of the ellipsoid onto this plane is

$$\frac{x_1 v_1^2}{a_1} = \frac{x_2 v_2^2}{a_2} = \frac{x_3 v_3^2}{a_3}.$$ (49)

By eliminating the quantities a_1, a_2, a_3. we obtain a relatively complicated equation for the ovaloid in terms of the coordinates x_1, x_2, and x_3. By transforming to polar coordinates we obtain the simple equation (47). The equation for the index ovaloid can be obtained in a similar way.

The derivation of the surfaces of two sheets is relatively complicated and will not be given here. Table II gives a comparison of all the eight optical surfaces described above.

Comparison between the Fresnel and Maxwell Theories. It was pointed out earlier that the main result of the electromagnetic theory for transparent isotropic media is the Maxwell law

$$n = \sqrt{\varepsilon}.$$

In order to apply this law to anisotropic media, it is necessary to consider whether the quantity n refers to the ray or the normal refractive indices (n_N or n_S). We shall attempt to solve this problem by comparing the dielectric constant surface and the Fresnel ellipsoid (46).

It is shown in books on crystal physics that the dielectric constant surface, i.e., the surface whose radius vectors are equal to the corresponding dielectric constant, is given by

$$\varepsilon = \varepsilon_1 c_1^2 + \varepsilon_2 c_2^2 + \varepsilon_3 c_3^2,$$ (50)

where ε is the dielectric constant which depends on direction in the crystal (the term "constant" is unfortunate since in this case the quan-

48

tity ε is in fact a variable), ε_1, ε_2, ε_3 are constants equal to the three principal dielectric constants, respectively. It is important to note that this surface is not an ellipsoid, although in the scientific literature it is often called and considered to be the dielectric constant ellipsoid.

TABLE II

Equations of Optical Surfaces

Simple surfaces	Surfaces of two sheets
Fresnel ellipsoid $$\frac{x_1{}^2}{v_1{}^2} + \frac{x_2{}^2}{v_2{}^2} + \frac{x_3{}^2}{v_3{}^2} = 1$$ $$\frac{1}{v_S{}^2} = \frac{c_1{}^2}{v_1{}^2} + \frac{c_2{}^2}{v_2{}^2} + \frac{c_3{}^2}{v_3{}^2}$$	Wave surface $$\frac{c_1{}^2 v_1{}^2}{v_S{}^2 - v_1{}^2} + \frac{c_2{}^2 v_2{}^2}{v_S{}^2 - v_2{}^2} + \frac{c_3{}^2 v_3{}^2}{v_S{}^2 - v_3{}^2} = 0$$
Velocity ovaloid $$v_N{}^2 = c_1{}^2 v_1{}^2 + c_2{}^2 v_2{}^2 + c_3{}^2 v_3{}^2$$	Normal velocity surface $$\frac{c_1{}^2}{v_N{}^2 - v_1{}^2} + \frac{c_2{}^2}{v_N{}^2 - v_2{}^2} + \frac{c_3{}^2}{v_N{}^2 - v_3{}^2} = 0$$
Optical indicatrix $$\frac{x_1{}^2}{n_1{}^2} + \frac{x_2{}^2}{n_2{}^2} + \frac{x_3{}^2}{n_3{}^2} = 1$$	Index surface $$\frac{c_1{}^2 n_1{}^2}{n^2 - n_1{}^2} + \frac{c_2{}^2 n_2{}^2}{n^2 - n_2{}^2} + \frac{c_3{}^2 n_3{}^2}{n^2 - n_3{}^2} = 0$$
Index ovaloid $$n_S^2 = c_1{}^2 n_1{}^2 + c_2{}^2 n_2{}^2 + c_3{}^2 n_3{}^2$$	$n_S \approx \dfrac{1}{v_S}$ surface $$\frac{c_1{}^2}{n_S^2 - n_1{}^2} + \frac{c_2{}^2}{n_S^2 - n_2{}^2} + \frac{c_3{}^2}{n_S^2 - n_3{}^2} = 0$$

Let us return to the equation of the Fresnel ellipsoid:

$$\frac{1}{v_S^2} = \frac{c_1^2}{v_1^2} + \frac{c_2^2}{v_2^2} + \frac{c_3^2}{v_3^2} .$$

Since the ray and normal velocities are equal along the principal axes of this ellipsoid, it follows that Maxwell's law may be used for these directions. Therefore, we can write

$$\varepsilon_1 = n_1^2 = \frac{c^2}{v_1^2} ; \quad \varepsilon_2 = n_2^2 = \frac{c^2}{v_2^2} ; \quad \varepsilon_3 = n_3^2 = \frac{c^2}{v_3^2} .$$

Substituting these quantities into (50) we have

$$\varepsilon = \frac{c^2 c_1^2}{v_1^2} + \frac{c^2 c_2^2}{v_2^2} + \frac{c^2 c_3^2}{v_3^2}.$$

A comparison of this equation with the equation of the Fresnel ellipsoid gives the following final result

$$\frac{1}{v_S^2} = \frac{\varepsilon}{c^2} \; ; \quad \varepsilon = \frac{c^2}{v_S^2} = n_S^2 \qquad (51)$$

or

$$n_S = \sqrt{\varepsilon}. \qquad (51')$$

Thus Maxwell's law $n = \sqrt{\varepsilon}$, when applied to anisotropic media, means that the square root of the dielectric constant is equal not to the refractive index $n = n_N$ (in the normally accepted sense) but the quantity n_S whose physical meaning (cf. p. 31) now becomes clear.

Returning now to Fresnel's original idea expressed by the equation

$$v_S = \sqrt{\frac{Q}{r}}, \qquad (52)$$

where Q is the elasticity and r is the density of the ether, we conclude from a comparison between this equation and equation (51) that from the point of view of the electromagnetic theory the elasticity of the ether is inversely proportional to the dielectric constant, so that

$$Q \approx \frac{1}{\varepsilon}.$$

This is the physical meaning of the imagined property of the ether in a crystalline medium.

Methods for Measuring Refractive Indices. In order to construct the above optical surfaces it is sufficient to know the three principal refractive indices of a crystal n_1, n_2, n_3. They are usually accurately measured by one of the following three methods: the prism method, the total internal reflection method and the immersion method.

1. The prism method. The crystal is made into a prism in such a way that the normal to the wave can be made accurately parallel to one of the principal axes of the indicatrix. The simplest way of doing this is to make one of the faces of the prism, say AB, a symmetry plane of the indicatrix so that the refracting edge A of the prism is parallel to one of the principal axes of the indicatrix (Fig. 48), and the incident beam of light is directed perpendicular to the face AB.

If the plane AB contains, for example, the n_3, n_2 axes of the indicatrix, then inside the prism two undeviated rays (they are also

50

normals) will be transmitted along the n_1 direction with the corresponding refractive indices n_3, n_2. By measuring the angles of devi-

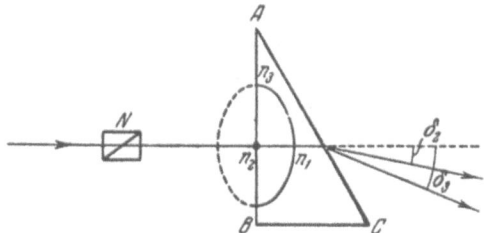

Fig. 48. Measurement of principal refractive indices by the prism method. A beam of light passes through the Nicol prism N and is incident on the face AB of the prism cut from the crystal. The face AB coincides with one of the symmetry planes of the indicatrix (in the drawing this plane coincides with the plane $n_2 n_3$). n_2, n_3 are determined by measuring the angles δ_2 and δ_3.

ation δ_2, δ_3 of these rays from the original direction, and knowing the angle A of the prism, the two refractive indices may be determined from the following formula

$$n_2 = \frac{\sin(A + \delta_2)}{\sin A}; \quad n_3 = \frac{\sin(A + \delta_3)}{\sin A}. \tag{53}$$

In order to measure n_1, one can either make a new prism in the appropriate way, or use the same prism with the incident beam perpendicular to the face BC, which should of course be cut parallel to the n_1, n_2 plane. In order to define the direction of vibrations in each of the deviated beams the incident light is first passed through a Nicol prism N.

2. The total internal reflection method is based on the use of a glass hemisphere whose refractive index is greater than that to be measured (Fig. 49). The crystal plate is cut parallel to one of the planes of symmetry of the indicatrix and is attached to the plane surface of the hemisphere by means of a drop of a liquid, whose refractive index should be greater than that to be measured. The plate is then rotated about its normal until one of the axes of the indicatrix which lies in the plane of the plate is in the plane of incidence (in the plane of the paper). Suppose, for example, that this axis is the n_1 axis. The only rays which can be transmitted along its direction are S_2, S_3 (they are also normals) with refractive indices n_2, n_3 which are to be measured.

When the incident beam transmitted by the hemisphere is incident normally on the crystal plate (angle of incidence equal to zero), it will

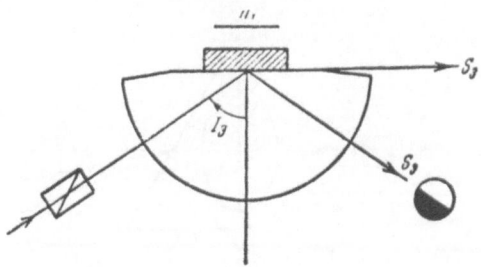

Fig. 49. Measurement of principal refractive indices by the total internal reflection method. Ray of light passes through the Nicol prism, enters a glass hemisphere and is reflected from a crystal plate which is attached to the hemisphere by means of a drop of liquid. The refractive index n_1 is determined by measuring the angle of total internal reflection I_3.

be split into two rays in the crystal plate but will emerge from it undeviated with respect to the incident ray. If the angle of incidence is gradually increased, we shall at first observe the transmission of both the rays through the plate and their different refraction on emerging from the plate. When the angle of incidence reaches some critical value I_3, the angle of refraction of one of the two rays will become equal to 90°, i.e., it will be transmitted in the direction of the n_1 axis. This will be the S_3 ray with refractive index n_3 smaller than n_2. When the angle I_3 is further increased the ray S_3 will remain totally internally reflected while the other ray will for some time be transmitted and refracted by the plate. At some angle I_2, greater than I_3, the second ray S_2 will become reflected also. The direction of this ray immediately before it becomes reflected is along the axis of the crystal. The corresponding refractive index is n_2.

If the relative refractive index for the ray S_2 passing from crystal to glass is denoted by n_{cg}, then in accordance with the law of total internal reflection (5) we have

$$n_{cg} = \frac{1}{\sin I_2}.$$

On the other hand, the law of refraction (3) gives

$$n_g = n_2 \cdot n_{cg}.$$

Hence

$$n_2 = n_g \sin I_2. \tag{54}$$

Similarly,

$$n_3 = n_g \sin I_3.$$

Thus, in order to determine the two principal refractive indices of a crystal, it is sufficient to measure two angles of total internal reflection I_2 and I_3, and to know the refractive index of the hemisphere. If the plate is sufficiently thick, and its other surfaces are parallel to the other symmetry planes of the optical indicatrix, the third principal refractive index n_1 can be determined by placing the plate on its side. Alternatively, a second plate can be prepared for this purpose.

In order to define the vibration directions, the incident or reflected rays are passed through a Nicol prism. An instrument including a glass hemisphere and used to measure the refractive indices of crystals is called a crystal refractometer.

3. The immersion method is based on the fact that a transparent body emersed in a liquid will seem to disappear when its refractive index is close to that of the liquid. The method is mainly used to measure the refractive indices of fine crystals and chips, but it can, of course, also be used for large crystals.

The determination of refractive indices of very fine objects is carried out under a microscope. An ordinary microscope can be used for optically isotropic crystals. In the case of optically anisotropic crystals one must use a polarizing microscope, i.e., a microscope equipped with a pair of Nicol prisms, one of which is located under the microscope stage and the other in the microscope tube.

In order to carry out the necessary measurement, one must have a collection of liquids with known refractive indices. The piece under investigation (we shall consider it, for the moment, to be optically isotropic) is placed in succession in different drops of these liquids until the disappearance effect is most pronounced. The refractive index of the crystal is then equal to the refractive index of the liquid.

The point of disappearance is best determined by observing the so-called Becke line. This is due to the combined effect of refraction and reflection at the boundary between the object and the liquid, and has the form of a luminous line following the outline of the object.

If the refractive index of the liquid is lower than that of the object under investigation, light will be deviated inward at the edges of the object. In this case the object will act similarly to a collecting lens.

As the tube of the microscope is raised the Becke line will move toward the center of the object. In the opposite case, it will move toward the liquid. When the Becke line disappears, the refractive indices of the liquid and the specimen should be equal. If for a given liquid the Becke line moves inward when the microscope tube is raised, and inward when another liquid is used, then the refractive index of the crystal is approximately equal to the mean of the indices of the two liquids.

The measurement of the principal refractive indices of optically anisotropic crystals is complicated by the fact that in each case the orientation of the crystal relative to the axis of the microscope must be known, and the fact that, in general, there are two refractive indices for each position of the crystal. The difficulties are resolved by the use of various rotating devices (Fedorov's table, Kolotushkan's pin, etc.) which may be used to orient the crystal in the required fashion, and the introduction of a Nicol prism into the microscope so that one of the rays can be extinguished while the refractive index for the other is being measured.

Indicatrix Dispersion. By the dispersion of light (in the widest sense) one normally understands the dependence of optical properties of media on the frequency of light. All forms of dispersion of light which are in some way associated with the phenomena of ordinary and double refraction in crystals may be reduced to the properties of the mutual disposition inside the crystal of indicatrices corresponding to different colors, i.e., different frequencies. This form of dispersion will be called indicatrix dispersion.

1. In cubic crystals indicatrices corresponding to different colors form a family of concentric spheres. This family of spheres clearly has the same symmetry as a single sphere ($\infty/\infty \cdot m$). Consequently, indicatrix dispersion does not introduce anything new into the optical symmetry of cubic crystals in comparison with the information obtained from the properties of a single monochromatic indicatrix.

In the cubic system, there are five crystal classes characterized by five groups of morphological symmetry, namely, $3/2$, $3/4$, $\bar{6}/2$, $3/\bar{4}$ and $\bar{6}/4$. Since the spherical symmetry group $\infty/\infty \cdot m$ includes all the possible symmetry elements of finite figures, one may consider that all the symmetry elements of the above groups are included in the spherical group, or in other words, each of the above groups is a subgroup of the $\infty/\infty \cdot m$ group.

In cubic crystals, and also in all other optically isotropic media, there are two forms of ordinary dispersion of light, i.e., two forms of dependence of refractive indices on frequency or wavelength in a vacuum, namely, normal and anomalous dispersion. Normal dispersion is characterized by the inequality

$$\frac{dn}{d\lambda} < 0; \tag{55}$$

and anomalous dispersion by

$$\frac{dn}{d\lambda} > 0. \tag{55'}$$

It is important to remember that a given body (not ideally transparent) can show anomalous dispersion in one part of the spectrum and normal dispersion in another.

2. In uniaxial crystals, the family of indicatrices corresponding to different colors forms a coaxial system of ellipsoids of revolution whose $m \cdot \infty : m$ symmetry does not differ from the symmetry of a single ellipsoid. It follows that, again, indicatrix dispersion in uniaxial crystals does not introduce into the optical symmetry of crystals anything new in comparison with the information given by a single monochromatic indicatrix.

One must note again that the ellipsoids of the family do not necessarily have to be all oblate or all prolate, and consequently, it is possible that the crystal can be positive for one wavelength and negative for another.

It was already mentioned that uniaxial crystals have a single special direction, i.e., they include all trigonal, tetragonal and hexagonal crystals. These crystals belong to the following nineteen groups of morphological symmetry.

3	—	$3 : m$	$3 \cdot m$	$3 : 2$	—	$m \cdot 3 : m$
4	$\overline{4}$	$4 : m$	$4 \cdot m$	$4 : 2$	$\overline{4} \cdot m$	$m \cdot 4 : m$
6	$\overline{6}$	$6 : m$	$6 \cdot m$	$6 : 2$	$\overline{6} \cdot m$	$m \cdot 6 : m$

It is immediately clear that all these groups are subgroups of the $m \cdot \infty : m$, group which includes the family of indicatrices of uniaxial crystals. This shows once again that the morphological symmetry group of a crystal either coincides with the group of the property under consideration or is a subgroup of the latter.

In connection with uniaxial crystals it is convenient to separate out the special case of birefringent dispersion from the general case of indicatrix dispersion. It is defined as the dependence of the birefringent power of the crystal on frequency or wavelength in a vacuum:

$$\Delta n = |n_o - n_e| = f(\nu) = \varphi(\lambda). \tag{56}$$

By the birefringent power of a crystal, or simply birefringence, we understand the absolute magnitude of the difference between the principal refractive indices $|n_o - n_e|$. The simplest and most accurate representation of this dependence is by means of a graph. Such a graph is shown in Fig. 50 for quartz, where Δn is plotted as a function of λ.

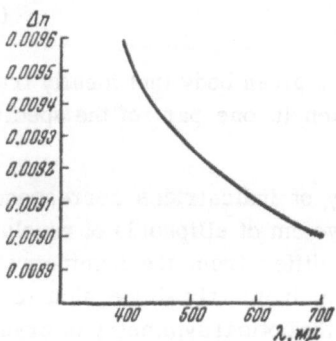

Fig. 50. Curve showing birefringent dispersion in quartz. The wavelengths in vacuum λ are plotted along the horizontal axis and the birefringent power (the difference between the principal refractive indices Δn) along the vertical axis.

In optical measurements on crystals, and especially in petrographic practice, birefringent dispersion can often be neglected and Δn can be looked upon as constant independent of wavelength:

$$\Delta n = \text{const.} \tag{57}$$

3. In orthorhombic crystals, indicatrices corresponding to different color are set with symmetry elements parallel to each other. It follows that the symmetry of the whole indicatrix family will coincide with the $m \cdot 2 : m$ symmetry of each separate indicatrix, the symmetry of the holohedral class of this system.

The orthorhombic system includes, as is well known, the following three groups of morphological symmetry: $m \cdot 2 : m$; $2 : 2$; $2 \cdot m$. The first of these coincides with the symmetry group of the indicatrix family and the remaining two groups are subgroups of this group. We see that the general crystallophysical law, formulated above for optically isotropic and for uniaxial crystals, holds in this case also.

In addition to ordinary dispersion and birefringent dispersion there is a third, new type of dispersion in orthorhombic crystals, namely, the dispersion of optic axial angles which is often called simply axial dispersion. Let us now briefly consider birefringent dispersion.

By birefringent dispersion in orthorhombic, and generally, biaxial crystals, one usually understands the dependence of the maximum difference between principal refractive indices on frequency or the wavelength in a vacuum, i.e.,

$$\Delta n = |n_3 - n_1| = f(\nu) = \varphi(\lambda). \tag{58}$$

In many cases the dependence of the quantities $|n_3 - n_2|$ and $|n_2 - n_1|$ on frequency or wavelength may be of interest. These three dependencies describe completely birefringent dispersion in biaxial crystals.

Let us now consider the dispersion of optic axial angles. In orthorhombic crystals, the angles between the optic axes of the indicatrices corresponding to different colors are different. In accordance with (43), this is explained by saying that in orthorhombic crystals the principal refractive indices n_1, n_2 and n_3 are different for different colors. On the other hand, the bisectrices of the optic axial angles in orthorhombic crystals are all coincident. This follows from the fact that the bisectrix of the optic axial angle of each indicatrix is also its twofold axis, and as was pointed out above, each symmetry element of any one indicatrix in othorhombic crystals coincides with corresponding symmetry element of any other indicatrix.

Two cases of dispersion of the optic axial angle are usually distinguished and are characterized by the inequalities

$$\left.\begin{aligned} V_v &> V_r, \\ V_v &< V_r, \end{aligned}\right\} \tag{59}$$

in which V_v and V_r are the optic axial half-angles for violet and red rays.

The magnitude of the dispersion of optic axial angles (or briefly axial dispersion) can in some crystals be quite considerable. For example, in rubidium sulfate crystals $2V = 28°30'$ for the red lithium line, and $2V = 41°35'$ for the blue F line.

Normally, the optic axes of the family of indicatrices lie in a single plane coinciding with one of the symmetry planes of the family, which is then called the optic axial plane (Fig. 51 a). Sometimes, and rather infrequently, one comes across crystals in which indicatrices corresponding to different colors are arranged in two symmetry planes of the family (Fig. 51 b). For example, in TiO_2 crystals the optic axes of the indicatrices lie in one symmetry plane for $\lambda = 550$ mμ and in another for $\lambda = 550$ mμ. For wavelengths accurately equal to 550 mμ the indicatrix is an ellipsoid of revolution with the optic axis lying along the line of intersection of the above two planes, i.e., coinciding with the twofold axis of the indicatrix family. At this wavelength brookite crystals are uniaxial and have a $m \cdot \infty : m$ symmetry. This example shows once again that a given crystal can have different symmetries for different properties.

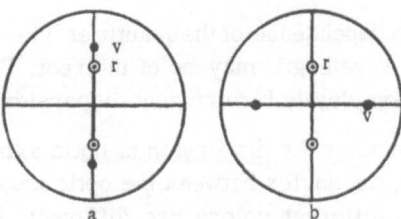

Fig. 51. Dispersion of optic axial angle. a) Normal case, V_v and V_r lie in the same plane; b) anomalous case, the planes of optic axes for red and violet rays cross. The black and white circles show the exits of the axes of violet and red indicatrices.

4. In monoclinic crystals the indicatrices of the family are arranged around a single, and common to them all, twofold axis, so that the axis of the entire aggregate of indicatrices should be of the same order (Fig. 52). The plane perpendicular to this axis is clearly a symmetry plane of the family. The presence of these two symmetry elements means that the indicatrix family has a center of symmetry. The symmetry group of the indicatrix family for any monoclinic crystal should therefore coincide with the $2 : m$ group of the holohedral monoclinic system.

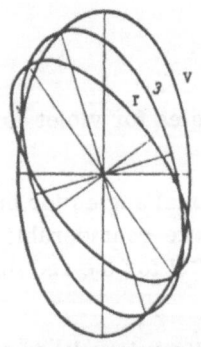

Fig. 52. Fan-shaped arrangement of indicatrices of different color around a common axis of symmetry in monoclinic crystals.

Crystals belonging to the $2 : m$, 2 and m groups of morphological symmetry belong to monoclinic crystals. The first of these groups coincides with the symmetry group of the indicatrix family and the other two are its subgroups. Thus, here too, the above crystallophysical rule is verified.

In addition to the above three forms of dispersion, which are characteristic of orthorhombic crystals, we have in monoclinic crystals a fourth form of dispersion which we shall call monoclinic dispersion of bisectrices.

Experimental methods of observation of the dispersion of bisectrices (and axes) in crystals will be described below (p. 105). Here we shall consider all the cases of monoclinic dispersion of bisectrices using a model consisting of a sphere whose surface shows the point of exit of bisectrices and optic axes of violet (v) and red (r) indicatrices.

If the observer looks at the model along the twofold axis of the indicatrix family, and if this axis is also the acute bisectrix for all

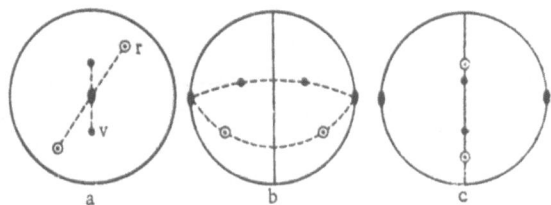

Fig. 53. Dispersion of axes and bisectrices in monoclinic crystals. a) Crossed dispersion; b) parallel dispersion; c) oblique dispersion.

the indicatrices of the family, then monoclinic dispersion of bisectrices will appear to him as crossed dispersion (Fig. 53 a). The optic axial planes of the indicatrices of the family will in this case all pass through the twofold axis at various angles to each other.

If the optic axial planes pass through the twofold axis as before, but the acute bisectrices lie in the plane of symmetry of the indicatrix family, and if the observer looks at the model along the plane of symmetry of the indicatrix family, the indicatrix dispersion will appear to him as parallel dispersion (Fig. 53 b).

The third case of the dispersion of bisectrices will occur when all the acute bisectrices lie in the plane of symmetry of the indicatrix family and the observer looks at the model along this plane. This case of dispersion of bisectrices is known as oblique dispersion (Fig. 53 c).

5. In crystals belonging to the triclinic system the dispersion of indicatrices is not subject to any symmetry requirements except that all the indicatrices of the family should have a common center of symmetry which should be the only symmetry element of the indicatrix family as a whole.

Crystals in the $\bar{2}$ and 1 morphological symmetry groups belong to the triclinic system. The first of these groups coincides with the symmetry group of the indicatrix family and the second is its subgroup. The crystallophysical rule frequently mentioned above is thus verified in this case also.

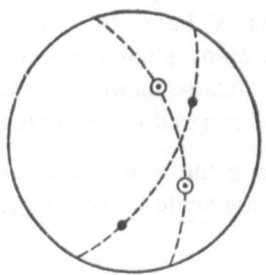

All the above forms of dispersion of indicatrices are possible in the case of crystals belonging to the triclinic system plus the skew dispersion of bisectrices which is illustrated in Fig. 54.

Fig. 54. General case of dispersion of axes and bisectrices in triclinic crystals.

Distribution of Crystals Among Five Double-Refraction Symmetry Groups. It is clear from the above discussion that all crystals belong to one of five optical symmetry groups, according to the character of indicatrix dispersion, or more generally, according to the character of their double refraction (Table III).

TABLE III

Distribution of Crystals Among Five Double-Refraction
Symmetry Groups

Double-refraction symmetry groups	$\bar{2}$	$2:m$	$m \cdot 2:m$	$m \cdot \infty : m$							$\infty/\infty \cdot m$	
Morphological symmetry groups of crystals	$\dfrac{1}{\bar{2}}$	2 m $2:m$	$2 \cdot m$ $2:2$ $m \cdot 2{:}m$	3 4 6	$\bar{4}$ $\bar{6}$	$3:m$ $4:m$ $6:m$	$3 \cdot m$ $4 \cdot m$ $6 \cdot m$	$3:2$ $4:2$ $6:2$	$\bar{4} \cdot m$ $\bar{6} \cdot m$	$m \cdot 3{:}m$ $m \cdot 4{:}m$ $m \cdot 6{:}m$	$3/2$ $3/4$	$\bar{6}/2$ $3/\bar{4}$ $\bar{6}/4$

It is important to note that two of these groups ($m \cdot \infty : m$ and $\infty / \infty \cdot m$) have infinite-fold symmetry axes. At first sight this looks strange since these axes are noncrystallographic and clearly contradict the lattice structure of crystals.

On the other hand, experimental evidence indicates that all the homogeneous noncrystalline substances such as water, liquid crystals, stretched rubber, etc., also belong to these five symmetry groups. It seems strange that under certain conditions (e.g., in the presence

60

of mechanical stresses, electric or magnetic fields, etc.) the optical symmetry of these media is described by the typically crystallographic groups $\bar{2}$, $2:m$, $m\cdot 2:m$.

These apparent contradictions can be explained away as follows. It is known that, within experimental errors, the thermal expansion coefficients of cubic crystals are independent of direction and hence, as far as the phenomenon of thermal expansion is concerned, cubic crystals may be looked upon as practically isotropic. Moreover, it can also be shown theoretically that such crystals must be isotropic with respect to thermal expansion since otherwise they could not have the characteristic lattice structure (the lattice would distort). This means that although generally crystals are anisotropic with respect to many of their properties, they can at the same time be isotropic with respect to some other properties. Thus the simultaneous isotropy and anisotropy of crystals cannot be considered as surprising.

In order to justify the optical isotropy of cubic crystals we must remember the following two factors. Firstly, it must be remembered that the wavelengths used in optics are a thousand times larger than the interatomic distances in the lattice, so that the latter cannot produce any diffraction effects in the visible part of the spectrum. Secondly, in material media, it is the weakly bound outer electrons which are responsible for optical phenomena. It is hardly correct to ascribe a lattice structure to this electron cloud, since the electrons in the cloud are in an uncorrelated, or only weakly correlated, motion. It follows that the electron cloud formed in a solid body out of the atomic outer electrons resembles a "texture" rather than a crystalline lattice. Strictly, the crystal lattice is formed only by the atomic nuclei of the crystal, but as was pointed out above these nuclei do not participate in optical phenomena. If we look upon the electron cloud formed by the optical electrons as a kind of "texture," we are led to the conclusion that the "texture" can have the same symmetry as the crystal, or a higher symmetry including the symmetry of the crystal itself, which is in fact confirmed by observation.

INTERFERENCE OF LIGHT IN CRYSTAL PLATES

Law of Malus (1810). Interference of light in crystal plates can be investigated with the help of polarizing instruments (orthoscopes, conoscopes, polarizing microscopes) which are equipped with two Nicol prisms or polaroids (Fig. 55). The light is incident from below and successively passes through the first Nicol prism (polarizer), the crystal plate and the second Nicol prism (analyzer). The Nicol prisms can be rotated about the vertical axis of the instrument, and the crystal plate can be rotated about both the horizontal and the vertical axis.

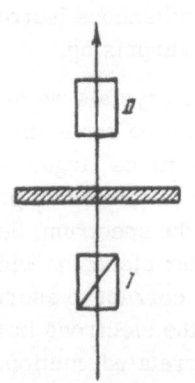

Fig. 55. Polarizing instrument used to study the interference of light in crystal plates. I and II are the lower and upper Nicol prisms in a crossed position; a crystal plate is placed between them.

Let us consider the intensity of the light transmitted by the instrument as a function of the relative angle of rotation between the two Nicol prisms in the absence of the crystal plate.

The intensity of polarized light transmitted by the first Nicol prism will be denoted by I_0. It is known that the energy of vibrations is proportional to the square of the amplitude, so that

$$I_0 = kA_0^2. \qquad (60)$$

We shall assume that the light travels toward the observer in a direction perpendicular to the plane of the paper (Fig. 56). The directions of vibrations transmitted by the first and second Nicol prisms are indicated by I and II, and are at an angle α to each other. Let us resolve the vector A_0 into two components, one of which is along II and the other along the line perpendicular to this direction and the plane of the paper. The first component is given by

$$A_1 = A_0 \cos \alpha,$$

and will be transmitted by the second Nicol prism while the second component is extinguished. The intensity I of the transmitted light should be proportional to the square of the amplitude A_1 and is given by

$$I = kA_1^2 = kA_0^2 \cos^2 \alpha = I_0 \cos^2 \alpha. \qquad (61)$$

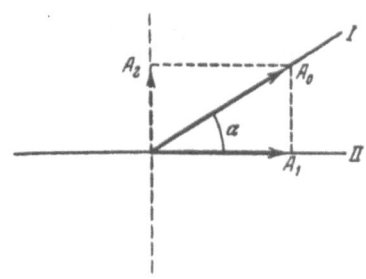

Fig. 56. Decomposition of the amplitude A_σ of waves leaving Nicol prism I into two components A_1, A_2 along and perpendicular to the vibration directions in Nicol prism II.

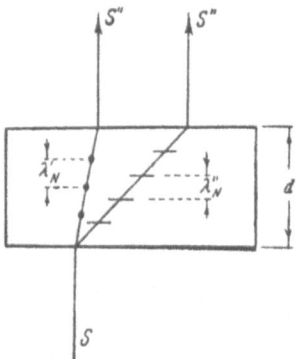

Fig. 57. Concerning the derivation of formula (64) for the path difference.

This equation expresses the law of Malus. It is, of course, strictly correct only for ideally transparent Nicol prisms, since it does not take into account the absorption of light in these prisms. When $\alpha = 0°$, i.e., in the case of parallel Nicol prisms, $I = I_0$ and the light is transmitted through the second prism without change in intensity. When $\alpha = 90°$, i.e., in the case of crossed Nicol prisms, $I = 0$ and the light is completely extinguished by the second Nicol prism.

Path Difference in a Crystal Plate. Consider a parallel beam S of natural or polarized light, incident normally on a crystal plate of thickness d (Fig. 57). We know that in the general case, i.e., in the case of a biaxial and arbitrarily oriented crystal, the incident beam of light should become decomposed inside the crystal into two beams traveling in different directions, and that the wave normals of the two beams should not be refracted under these conditions (cf. Fig. 27).

Let λ_N' and λ_N'' denote the wavelengths along the normals in the two beams inside the crystal. It follows that the number of wavelengths along the path of the two beams inside the crystal is

$$\frac{d}{\lambda_N'} \text{ and } \frac{d}{\lambda_N''}$$

63

respectively. It follows that, inside the plate, one of the beams leads the other by the number of wavelengths given by the expression

$$G = d \left| \frac{1}{\lambda'_N} - \frac{1}{\lambda''_N} \right|. \tag{62}$$

When the two beams leave the plate, this path difference will be preserved, but because their wavelengths are now equal to, say, λ, the path difference Γ, i.e., the shift of one system of waves relative to the other, is now expressed not in terms of the number of waves but in the units of length, and is given by

$$\Gamma = \lambda G = d \left| \frac{\lambda}{\lambda'_N} - \frac{\lambda}{\lambda''_N} \right| \tag{63}$$

or

$$\Gamma = d \, | n' - n'' |. \tag{64}$$

The path difference is usually expressed in millionth parts of a millimeter (mμ) or in angstroms (Å). The thickness of the plate should be expressed in the same units. The quantity

$$G = \frac{\Gamma}{\lambda}, \tag{65}$$

which gives the path difference in terms of the number of wavelengths, will be called the wave path difference, to distinguish it from Γ. The quantity $|n' - n''|$ will be called the birefringent power of the plate, which should not be confused with the birefringent power of the crystal, i.e., the difference between the maximum and minimum refractive indices (p. 54).

We have so far considered only the case where a narrow beam of light is doubly refracted in a thick plate. In this case, the beam is split into two parts inside the crystal and the two beams are parallel when they emerge from the crystal although they do not travel along the same path. It is easy to see that if one takes a sufficiently wide incident beam, and a sufficiently thin plate, one can always arrange for the two beams emerging from the crystal to travel along the same path and almost completely overlap. In this case, one has to deal with interference phenomena produced by means of crystal plates.

Amplitude of Interfering Vibrations. Consider two vectors A_1 and A_2 which rotate with the same angular velocity and in the same direction, but have different phases Δ_1 and Δ_2 (Fig. 58). The phase difference must, of course, remain constant and we shall denote it by Δ. We know that a uniform rotation of a vector is associated with harmonic vibrations of the component along one of the diameters of the circle. Such vibrations will be executed by the vectors a_1 and a_2, which are the projections of A_1 and A_2 on the vertical diameter of the circle.

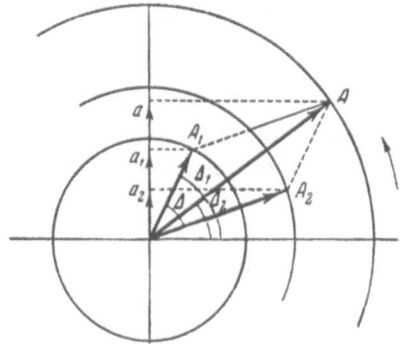

Fig. 58. Drawing illustrating the derivation of the formula for the amplitude of interfering vibrations (67). Vectors A_1, A_2, and their resultant A, rotate about a common center with the same angular velocity. Their projections a, a_1, a_2 execute harmonic vibrations with amplitudes A, A_1, A_2. The phase difference Δ remains constant.

The amplitudes of vibration of the vectors a_1 and a_2 are equal to A_1 and A_2, respectively. The resultant of the two vectors A_1 and A_2 is equal to the diagonal A of the corresponding parallelogram. Since the vector A is rigidly connected to the vectors A_1 and A_2 it follows that when A_1 and A_2 rotate uniformly, the vector A will also rotate with the same angular velocity as A_1 and A_2.

This means that the projection a of the vector A should execute harmonic vibrations with an amplitude A. This is, in fact, the amplitude of the interfering vibrations of the vectors a_1 and a_2.

We know from trigonometry that the square of the diagonal of a parallelogram can be expressed in terms of its sides and the angle between them in the following way:

$$A^2 = A_1^2 + A_2^2 + 2A_1A_2 \cos \Delta. \qquad (66)$$

This formula may be rewritten in another form if one remembers that the following relation holds:

$$\cos \Delta = 1 - 2\sin^2 \frac{\Delta}{2}.$$

Substituting this expression into (66) we have

$$A^2 = (A_1 + A_2)^2 - 4A_1A_2 \sin^2 \frac{\Delta}{2}.$$ (67)

This formula may be used to calculate the resultant amplitude A when the component amplitudes A_1 ánd A_2 and the constant phase difference Δ are known.

This phase difference can, if necessary, be expressed in terms of $\Gamma = d(n'-n'')$ in the following way. It is known that if the path difference Γ is equal to λ, then the phase difference Δ is equal to 2π. This means that

$$\frac{\Gamma}{\lambda} = \frac{\Delta}{2\pi}.$$

It follows that

$$\Delta = 2\pi\frac{\Gamma}{\lambda} = 2\pi\frac{d(n'-n'')}{\lambda}.$$ (68)

The Transmission Formula. If a parallel beam of light is successively passed through a Nicol prism, a sufficiently thin crystal plate and a second Nicol prism, then all the conditions for interference to

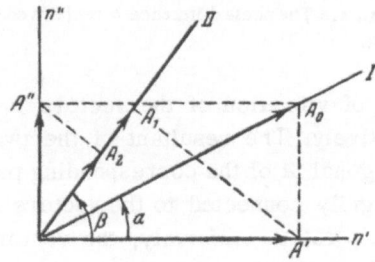

Fig. 59. Concerning the derivation of the transmission formula (73).

take place are satisfied, namely, (1) the two interfering rays are coherent, (2) their frequencies are equal, (3) there is a path difference and (4) their vibrations lie in the same plane. In fact, each monochromatic ray in a beam is split into two rays of equal frequency inside the crystal plate, and the two rays are coherent since they originate from the same source. Since they are transmitted through the plate with different velocities, there will in general be a path difference between them when they emerge from the plate. The second Nicol prism ensures that their vibration directions are in the same plane.

The above discussion may give the impression that observable interference phenomena could take place without the first Nicol prism. In fact, it will be shown below (p. 90) that it is essential.

Consider a monochromatic ray of light perpendicular to the plane of the paper (Fig. 59) which has successively passed through the first Nicol prism, the crystal plate and the second Nicol prism. The ray leaves the first prism with vibrations along the line I, and can be resolved into two linearly polarized rays with vibrations along mutually perpendicular directions n' and n'' parallel to the axes of the section of the indicatrix.

If the amplitude of the ray leaving the first Nicol prism is A_0, then the amplitudes of the component rays will as before be given by

$$A' = A_0 \cos\alpha; \quad A'' = A_0 \sin\alpha,$$

where α is the angle between the n' and I directions.

On entering the second Nicol prism each of these rays will again change their amplitude according to the same law. As a result, rays leaving the second Nicol will have amplitudes

$$A_1 = A_0 \cos\alpha \cos\beta; \quad A_2 = A_0 \sin\alpha \sin\beta, \tag{69}$$

where β is the angle between the n' and II directions.

Since the frequency of the two rays transmitted by the second Nicol prism is the same, and since their vibrations are in the same plane, we can use formula (67) given in the previous paragraph:

$$A^2 = (A_1 + A_2)^2 - 4A_1 A_2 \sin^2 \frac{\Delta}{2}.$$

Substituting the values of A_1 and A_2 obtained above into (69) we have

$$A^2 = (A_0 \cos\alpha \cos\beta + A_0 \sin\alpha \sin\beta)^2 -$$
$$- 4A_0^2 \sin\alpha \sin\beta \, \cos\alpha \cos\beta \sin^2 \frac{\Delta}{2}.$$

Using the trigonometric formula

$$\cos\alpha \cos\beta + \sin\alpha \sin\beta = \cos(\beta - \alpha),$$
$$2 \sin\alpha \cos\alpha = \sin 2\alpha,$$

we have, after substitution,

$$A^2 = A_0^2 \cos^2(\beta - \alpha) - A_0^2 \sin 2\alpha \sin 2\beta \sin^2 \frac{\Delta}{2}. \tag{70}$$

This formula may be used to calculate the amplitude A of rays passed through the second Nicol prism if the original amplitude A_0, the angles α, β and the phase difference Δ are known.

If we replace A^2 and A_0^2 by the intensities I and I_0 which are proportional to them, the above formula may be written in the following form

$$I = I_0 \cos^2 (\beta - \alpha) - I_0 \sin 2\alpha \sin 2\beta \sin^2 \frac{\Delta}{2} . \qquad (71)$$

Since it is often more important to know the fraction

$$\frac{I}{I_0} = J,$$

(the transmission) rather than the intensity I itself, the above formula may be rewritten in the form

$$J = \cos^2 (\beta - \alpha) - \sin 2\alpha \sin 2\beta \sin^2 \frac{\Delta}{2} . \qquad (72)$$

Substituting the expression for Δ given by (68), we finally have

$$J = \cos^2 (\beta - \alpha) - \sin 2\alpha \sin 2\beta \sin^2 \frac{\pi d \, | \, n' - n'' \, |}{\lambda} . \qquad (73)$$

This is the transmission formula.

In the following discussion we shall contrast transmission with absorption, i.e., the relative fraction of absorbed, or more accurately, untransmitted light (for example, light deflected to the side by a Nicol prism). The fractions of transmitted and absorbed light clearly add up to unity:

$$J + J_{abs} = 1. \qquad (73')$$

Some Ideas from Colorimetry. Before we pass on to the discussion of the transmission formula and the study of interference hues in crystal plates, it is necessary to recall some experimental data from colorimetry.

The human eye can directly perceive, and to some extent quantitatively estimate, the following three main properties of light: intensity, hue and hue density. The color of light is determined by the last two properties. White light has no hue and only one of the above three properties, namely, intensity. Black "light" (darkness) has zero intensity. We can assign to it any hue we please or look upon it as hueless, since if we reduce the intensity of white or any colored light to zero, we shall in all cases reach darkness. A reduction in the intensity of light can therefore be looked upon as an addition of darkness or black "light". Saturated hue is a hue of maximum density. Desaturated hue may be looked upon as saturated hue "diluted" with white light.

Each of the above three properties can, in the first rough approximation which is quite sufficient for our purposes, be characterized by a physical quantity: brightness by energy, hue by the mean wavelength and hue density by the ratio of the amount of light (energy) of saturated hue to the total amount of white desaturated light.

For our purposes, the discussion of interference phenomena in crystal plates may be based on the following three considerations.

Firstly, the color of light is uniquely determined by its spectrum, i.e., the intensity distribution curve with frequency or wavelength in vacuo. The converse of this is not true, since two equal colors can exist which do not have identical spectra.

Secondly, when two equally directed beams of light are mixed their spectra are added arithmetically, i.e., the ordinates of the one are simply added to those of the other.

Thirdly, we must consider the physiological law of color mixing. Before we formulate this law we must consider the following concepts.

Let us divide the solar spectrum into a number of intervals, say eight. Each of the intervals will be given a definite average color and a corresponding name will be given to it. We shall thus obtain a scale of principal colors, for example

$$r, o, y, y\text{-}g, g, l\text{-}b, d\text{-}b, v$$

(red, orange, yellow, yellow-green, green, light blue, dark blue, and violet).

By arranging these colors in the spectral order in sectors of a circle, we obtain a color rosette (Fig. 60 opposite p. 90). Let us agree to call the direction of the arrows as the direction of increasing color and the opposite direction as the direction of decreasing color. Colors of opposite sectors in the color rosette will be called complementary colors. It is often convenient to talk about colors intermediate between the adjacent or more distant colors in the color rosette. Such intermediate colors can be given compound names, e.g., orange-red, red-violet (purple), etc.

We can now give the following formulation of the law of color mixing: when two colors (equal intensity and equal hue density) are mixed, one obtains an intermediate color which is the more saturated the nearer are the two colors, while maximally distant, and therefore complementary, colors have no hue on mixing (white light). By color mixing we shall always understand the mixing of two beams of light of different hue but traveling in the same direction, and not the mixing of pigments which will not be of interest to us. The mixing of a number of colors can always be reduced to successive mixing of the colors in pairs.

The concept of complementary spectrum plays an important role in many colorimetric problems. Two spectra whose sum gives another spectrum are said to be mutually complementary with respect to it. When the third spectrum is not specified, two complementary spectra are understood to give the solar spectrum on addition.

It is important to remember that the complementarity of two spectra with respect to the white solar spectrum does not uniquely determine the hues of the two spectra, since the white solar spectrum can be obtained by mixing two white beams, a white beam and darkness, a red beam and a green beam, etc. Clearly, transmission and absorption spectra are always complementary to each other with respect to the spectrum of the primary light.

The above color nomenclature does not include such commonly employed terms as grey, brown, rose, etc. colors. These colors belong to the intermediate category. We have already discussed colors intermediate in hue. The purple color mentioned above is an example. Grey color is one of the colors intermediate in intensity and may be looked upon as white light of reduced intensity. Similarly, brown light may be looked upon as dark red, i.e., red color of reduced intensity. Rose color belongs to colors intermediate in hue density, and is intermediate between white and red.

General Remarks on the Transmission Formula. The transmission formula (73) derived above, contains five independent variables, namely, α, β, d, $(n' - n'')$, and λ. Our problem includes the determination of the role of each of these in the formation of the interference hue when the first Nicol prism transmits white light.

Let us write the transmission formula in the form

$$J = B + C, \tag{74}$$

where

$$B = \cos^2 (\beta - \alpha), \tag{75}$$

$$C = - \sin 2\alpha \sin 2\beta \sin^2 \frac{\pi d (n' - n'')}{\lambda}. \tag{76}$$

We note that B does not contain λ. This means that if we use white light in the observation of the interference hue, each of its monochromatic components transmitted by the polarizing instrument contributes the same fraction B of the (unit) primary intensity. In other words, B describes the uncolored (white) part of the transmission, and hence the quantity B is called the white component.

The white component is always positive since it is equal to the square of the cosine of $(\beta - \alpha)$. This means that it always determines a fraction of transmitted energy and not the energy absorbed by the instrument.

The second term C contains λ and hence describes the fraction of energy associated with the colored light. The term C is therefore called the color component. Since this component can take up any value between -1 and $+1$ it can describe either transmission of colored light ($C > 0$) or its absorption ($C < 0$). If the component C is zero, the transmitted light is purely white, while if the component B is zero, the transmitted light can have a dense hue. In the general case, the transmitted light has a hue intermediate between white and the most dense hue. A change in the sign of the color component indicates a transition from a given spectrum of the transmitted light to the complementary spectrum, since in the one case light of a given hue is added to white light, while in the other the same light is subtracted from white light.

Dependence of Transmission on Wavelength. It could be concluded from the above discussion that the transmission of white light through a doubly refracting crystal plate placed between two Nicol prisms leads to the appearance of interference colors. This is due to the fact that monochromatic waves which make up the white light have different transmissions. In order to determine which color should be assumed by a given crystal plate, it is necessary to investigate the dependence of transmission on wavelength.

Since all the quantities other than J and λ which enter into the transmission formula are conventionally considered to be constant, this formula can be written in the following form:

$$J = B + a \sin^2 \frac{b}{\lambda}, \qquad (77)$$

where B, a, and b are constants.

If J is plotted along the vertical and λ along the horizontal axis, then equation (77) gives a periodic curve with unequal periods which increase in the positive direction of the horizontal axis. The entire graph lies in the positive quadrant since J and λ are essentially positive.

If instead of λ one plots the wave number $k = \frac{1}{\lambda}$, along the horizontal axis, then the resulting curve is periodic, has equal periods (Fig. 61), and lies entirely above the B line corresponding to the white component. The interference color is determined only by that part of the area under the curve which corresponds to the visible part of the spectrum, i.e., the region between $k = \frac{1}{800}$ and $k = \frac{1}{400}$ mμ^{-1}. One can obtain information on the spectral composition of the color component and the magnitude of the white component, by dividing the area

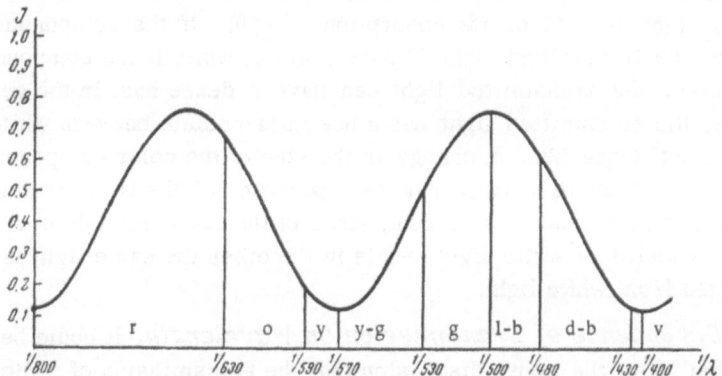

Fig. 61. Graph of the transmission formula (73). Dependence of the transmission J on $\frac{1}{\lambda}$ for $\alpha = 20°$, $\beta = 50°$, $\pi d(n'-n'') = 360000$. λ and d are expressed in mμ.

under the graph, which lies between the curve and the level corresponding to the value of B into seven intervals, and coloring each of them with the appropriate color in accordance with the adopted color rosette, while the area below the B level remains white.

It is easy to see that the number of periods of the curve which fall in this interval may differ depending on the value of the quantity

$$b = \pi d \, |n' - n''|,$$

which is always positive. The smaller the value of this quantity, the smaller the number of periods of the curve which can be fitted into the above interval. When the value of b is very small, only a fraction of a period falls in it. When the value of b is large, a correspondingly large number of periods falls into the above interval.

72

We shall illustrate with an example, the approximate calculation of the interference color of a crystal plate under concrete conditions.

Suppose we are given a quartz plate cut parallel to the optic axis. The thickness of the plate is $d = 0.1$ mm and $n' - n'' = 0.009$. The plate is placed between crossed Nicol prisms $\left(\beta - \alpha = \dfrac{\pi}{2}\right)$ in a diagonal position $\left(\alpha = \dfrac{\pi}{4}\right)$. Substituting these values into the transmission formula (73) we obtain

$$J = \sin^2 \frac{2826}{\lambda},$$

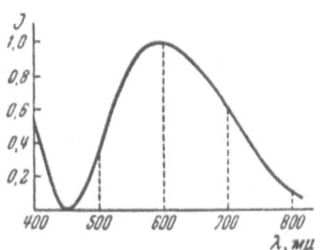

Fig. 62. Graph of the transmission formula (73) for a quartz plate cut parallel to the optic axis, placed between crossed Nicols and 0.1 mm thick. The dependence of the transmission J on λ is shown.

if the thickness is expressed in mμ. The graph of this formula (Fig. 62) has a maximum at $\lambda = 600$ mμ, i.e., in the orange part of the spectrum, and a minimum at $\lambda = 450$ mμ, i.e., in the blue. The white component is absent under these conditions. Using the adopted color rosette and the rules for using it which were described earlier, we find that our plate will have a dense orange color.

Dependence of Transmission on the Angle β. A change in the value of β is equivalent to a rotation of the second Nicol prism. We must therefore investigate the change in the interference color of the crystal plate when the second Nicol prism is rotated about the axis of the polarizing instrument.

It is clear that when the second Nicol prism is given one complete rotation, the white component

$$B = \cos^2 (\beta - \alpha)$$

passes twice through zero and twice through a maximum. The first will occur when the Nicol prisms are crossed, i.e., when $\beta = \alpha + \dfrac{\pi}{2}$ and $\beta = \alpha + \dfrac{3\pi}{2}$, and the second when the Nicols are parallel, i.e., when $\beta = \alpha$ and $\beta = \alpha + \pi$. For all intermediate values of β, the white component will have intermediate and therefore always positive values.

Since we have agreed that only β, is varied and all the other quantities in the transmission formula remain constant, the color component may be represented by

$$C = -a \sin 2\beta, \tag{78}$$

where a is a constant which is different for different wavelengths. When the second Nicol prism is fully rotated under these conditions, the color component should pass four times through the zero value, and four times through a maximum or minimum. It will be equal to zero when $\beta = 0, \frac{\pi}{2}, \pi,$ and $\frac{3\pi}{2}$, i.e., when the vibration direction in the second Nicol prism coincides with the vibration direction in the plate (one of the vibration directions in the plate parallel to the vibration direction in the second Nicol prism). The color component will pass through a minimum (it will be equal to $-a$) when $\beta = \frac{\pi}{4}, \frac{5\pi}{4}$ and through a maximum $(+a)$ when $\beta = \frac{3\pi}{4}, \frac{7\pi}{4}$. These four values of β, correspond to a diagonal position of the plate with respect to the position of the rotated Nicol prism.

The dependence of each of the components on the angle β may be represented graphically by plotting the angles β along the horizontal

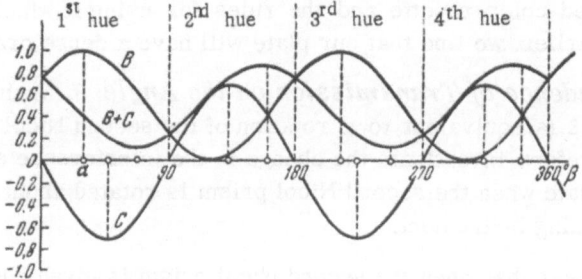

Fig. 63. Graph of the transmission formula (73). Dependence of the transmission $J = B + C$ on the angle of rotation β of the second Nicol prism (Fig. 59). B is the white component, C the color component, and $B + C$ their sum. The graph has been plotted assuming $\alpha = 30°$ and $k = 0.7$.

axis and the values of B and C along the vertical axis (Fig. 63). When the ordinates of the two curves are added algebraically, one obtains a curve representing the transmission $J_\beta = B + C$ as a function of the angle β. It is clear from Fig. 63 that J_β does not become zero at any value of β but has two maxima and two minima for β between 0 and 2π.

Summing up, we may say that the analysis of the dependence of J_β on β gives sixteen special points on the horizontal axis of the above diagram, and hence sixteen special positions of the second Nicol when it is fully rotated about the axis of the polarizing instrument. These points are the following.

1. Four points at which the C curve cuts the horizontal axis.
2. Four points corresponding to the extremal values of C (two maxima and two minima).
3. Four points corresponding to the four extremal values of B.
4. Four points corresponding to the extremal values of $J_\beta = B + C$.

An optical effect corresponds to each of these points in the interval between $\beta = 0$ and $\beta = \pi$. Of the eight such effects (which repeat themselves when the Nicol prism is rotated through 180°) the unaided eye can distinguish (although not always sufficiently clearly) only the following: 1) relative extinction and clearing of the plate corresponding to the minima and maxima of the quantity J_β, 2) transitions from more saturated hue to less saturated hue at the maxima and minima of the C curve, and 3) transitions from one hue to another, complementary, hue through bright white or grey light at the points of intersection of the C curve with the horizontal axis.

In connection with the transition from a given color to the complementary color it should be noted that this is exactly obtained when the upper Nicol prism is rotated through an angle of $\frac{\pi}{2}$ from any original position. This statement may be proved mathematically as follows.

Let J_β denote the transmission in the original position of the Nicol prism, i.e., for given angle β, and $J_{\beta+\frac{\pi}{2}}$ the corresponding value when the Nicol prism is rotated through $\frac{\pi}{2}$. In accordance with formulas (73) and (77) we may write

$$J_\beta = \cos^2(\beta - \alpha) - a\sin 2\beta,$$

$$J_{\beta+\frac{\pi}{2}} = \sin^2(\beta - \alpha) + a\sin 2\beta.$$

On adding these equations together we obtain

$$J_\beta + J_{\beta+\frac{\pi}{2}} = 1. \tag{79}$$

A comparison of the latter equation with equation (73') which was obtained earlier, leads to the conclusion that the transmission in the first position of the Nicol prism is equal to the absorption in the second, and conversely, and hence the transmission spectrum in the first case is complementary to the transmission spectrum in the second.

We have discussed the general case of the dependence of the transmission on the angle β. In practice however, one almost invariably deals with the special case α = 45°, i.e., when the crystal plate is in a diagonal position with respect to the fixed Nicol prism.

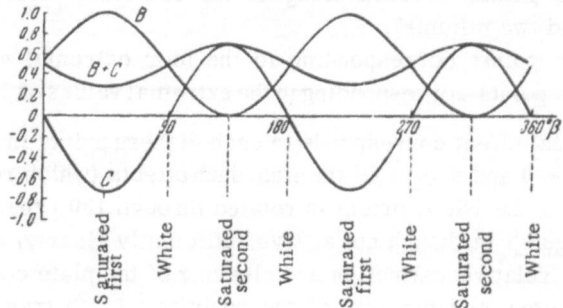

Fig. 64. Graph of the transmission formula (73). Dependence of the transmission J on the angle β for α = 45° and k = 0.7.

The corresponding diagram is shown in Fig. 64. We see that in this special case, the special points of the B, C and $B+C$ curves are accurately one above one other so that instead of sixteen special positions of the rotated Nicol prism we have only eight. At four of these points $\left(\beta = 0, \frac{\pi}{2}, \pi, \frac{3\pi}{2}\right)$ where the vibration direction in the rotated Nicol prism coincides with one of the vibration directions in the plate, the color component is absent and the plate becomes white. In all other positions of the second Nicol the plate becomes colored. The hue is most dense at four positions of the Nicol prism corresponding to the extremal values of C $\left(\beta = \frac{\pi}{4}, \frac{3\pi}{4}, \frac{5\pi}{4}, \text{and } \frac{7\pi}{4}\right)$. At the same time, in the intervals β = 0 to β = $\frac{\pi}{2}$ and β = π to β = $\frac{3\pi}{2}$ one hue is observed, and in the two other intervals, i.e., β = $\frac{\pi}{2}$ to β = π and β = $\frac{3\pi}{2}$ to β,= 2π, the complementary hue is observed. The transmission $J_\beta = B + C$ assumes maximum and minimum values at those values of β for which the plate has the densest hue. For all other positions of the second Nicol prism, the plate has an intermediate hue.

76

Thus when the second Nicol prism is given a complete rotation, the plate will change its saturated hue into the complementary one (according to the spectral composition) four times, when it passes through the less saturated hues, and four times when it passes through white light.

So far we have assumed that only the upper Nicol prism is rotated. However, it is easy to verify that all the above conclusions still hold when the upper Nicol prism remains fixed (β = const) and the lower Nicol prism is rotated (α varied).

Dependence of Transmission on the Rotation of the Plate. A rotation of the plate about the axis of the instrument with the two Nicol prisms kept fixed is equivalent to a simultaneous variation in β and α with ($\beta - \alpha$) kept constant.

Under these conditions, the white component will remain constant:

$$B = \cos^2(\beta - \alpha) = a,$$

and the color component which can now be represented by

$$C = -b \sin 2\alpha \sin 2\beta$$

will change periodically so that when the plate is given a complete rotation it will pass eight times through zero, four times through a minimum, and four times through a maximum. In this formula, the quantity b is a constant which is different for different wavelengths. It follows that in this case, the transmission may be written down in the form

$$J = a - b \sin 2\alpha \sin 2\beta. \tag{80}$$

The dependence of the quantities B, C, J on the angle of rotation of the plate may be represented by the diagram shown in Fig. 65 which

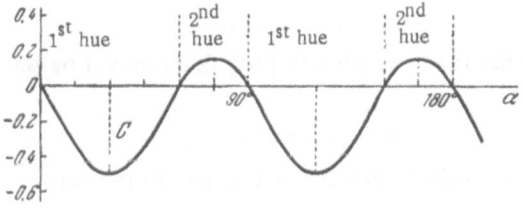

Fig. 65. Dependence of the color component C on the angle α for $\beta - \alpha = 30°$.

was drawn for the case $\beta - \alpha = 30°$. It is clear from this diagram that the color component is absent, i.e., the plate becomes white, when α = 60, 90, 150, 180°, and so on. At intermediate positions one observes one of two hues: the first hue will correspond to negative values of the color component C, and the second, complementary, hue will correspond to positive values of C. It is interesting to note that the intervals of α corresponding to these two hues are not equal. In our example one of the intervals is equal to 60° and the other to 30°. The most dense hue is obtained near the minima and maxima of the color component. The maxima (minima) of the color component also correspond to the maxima (minima) of the transmission. It is important to note that in the general case, i.e., when the angle (β — α) between the vibration directions in the Nicol prisms is arbitrary, the plate is never fully extinguished for any values of α, and its maximum brightness is obtained not when it is white but when it assumes one of the most dense hues.

In practice, the most interesting special cases are those when the Nicol prisms are crossed (×) and when they are parallel (∥).

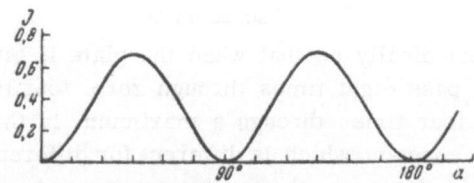

Fig. 66. Graph of the transmission formula (73) for a plate rotated between first Nicol prisms. Dependence of J on α for β − α = 90°.

When the Nicol prisms are crossed (β — α = 90°) the white component is absent, i.e.,

$$B = \cos^2 (\beta - \alpha) = 0.$$

It follows that the transmission is fully determined by the color component:

$$J_\times = - a \sin 2\alpha \sin 2\beta,$$

which in this case may be put into the very simple form

$$J_\times = a \sin^2 2\alpha, \tag{81}$$

since $\sin 2\beta = \sin 2 (\alpha + 90°) = - \sin 2\alpha$.

It is clear from this formula that when the Nicol prisms are crossed and the plate is given a complete rotation, the transmission is zero at four points ($\alpha = 0, \dfrac{\pi}{2}, \pi, \dfrac{3\pi}{2}$) and a maximum at four points ($\alpha = \dfrac{\pi}{4}, \dfrac{3\pi}{4}, \dfrac{5\pi}{4}, \dfrac{7\pi}{4}$) (Fig. 66). The plate will be extinguished in those positions for which $J_\times = 0$. In all other positions it will have a maximally dense and constant hue since the sign of the color component (+) remains the same. The plate will have a maximum brightness at the four positions which correspond to $J_\times =$ max. Altogether, when the plate is fully rotated it will be fully extinguished four times and fully translucent four times without changing the character of its hue in the intermediate positions, the only changes being those in the brightness.

When the Nicol prisms are parallel ($\beta - \alpha = 0$) the transmission formula assumes the form

$$J_\parallel = 1 - a \sin^2 2\alpha. \tag{82}$$

We saw earlier (79) that the transition from crossed to parallel Nicol prisms always leads to the complementary hue (according to the spectrum). This also occurs in the present case since

$$J_\times + J_\parallel = 1.$$

Thus when the plate is located between parallel prisms its hue in each position will be complementary to the corresponding hue between crossed Nicol prisms. When the plate is given a complete rotation it will pass through a white position four times $\left(\alpha = 0, \dfrac{\pi}{2}, \pi, \dfrac{3\pi}{2}\right)$ and will assume a hue in the intermediate positions. The character of the hue will remain constant and only its density will change, since in the case of parallel Nicol prisms, as opposed to crossed prisms, the white component is present. The most dense color will occur at $\alpha = \dfrac{\pi}{4}, \dfrac{3\pi}{4}, \dfrac{5\pi}{4}, \dfrac{7\pi}{4}$, i.e., when the plate is in the position of minimum brightness, and the plate will be white when the plate is in the position of maximum brightness.

Dependence of Transmission on the Thickness of the Plate.

This dependence may be investigated by placing thin and equally oriented crystal plates one upon the other, for example, mica plates cut along the cleavage planes, and studying the piles of plates thus obtained in a polarizing instrument. An even simpler method is to use a single plate having a variable thickness, i.e., a wedge. A quartz wedge cut so that one of the wide faces is parallel to the optic axis is most commonly used. The long edge of this face can be either parallel to the optic axis or perpendicular to it. We shall assume that this edge

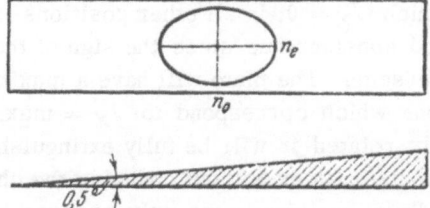

Fig. 67. A quartz wedge in plan and elevation.
The wedge is cut parallel to the optic axis which
is parallel to the length of the wedge.

is parallel to the optic axis, i.e., the major axis n_e of the quartz indicatrix (Fig. 67). The maximum thickness of the wedge is approximately 0.5 mm, its length of the order of 5 cm, and its angle about 0.5°. In order to prevent breakages, the wedge is usually covered by glass plates cemented to it with the aid of Canada balsam.

The dependence of J on d, the other quantities which enter into the transmission formula being constant, is represented by a periodic curve which is very similar to the curve giving the dependence of J on the wave number $k = \frac{1}{\lambda}$, (Fig. 62), since the two quantities d and $\frac{1}{\lambda}$ enter into the formula

$$\frac{\Delta}{2} = \frac{\pi d \mid n' - n'' \mid}{\lambda}$$

in a similar way.

Let us consider the dependence of J on d for the most common case, i.e., when the quartz wedge is placed in a diagonal position ($\alpha = \frac{\pi}{4}$) between crossed Nicols ($\beta - \alpha = \frac{\pi}{2}$). Under these conditions the transmission formula assumes the following simple form $n_e - n_o$

$$J = \sin^2 \frac{\pi d (n_e - n_o)}{\lambda} \tag{83}$$

Substituting the appropriate value for π and the known value of which for quartz is equal to 0.009, we obtain

$$J = \sin^2 0.02826 \frac{d}{\lambda} \tag{84}$$

Figure 68[*] shows curves of the dependence of J on d for five different values of λ (in order to simplify the drawing only five colors

[*]Figure 68 will be found opposite page 90.

were used instead of the eight in the adopted color rosette). It follows from these curves that when the quartz wedge is observed between crossed Nicol prisms in a parallel beam of monochromatic light, one should see a series of dark and bright bands which continuously merge into each other (Fig. 69). It may therefore be concluded that a quartz

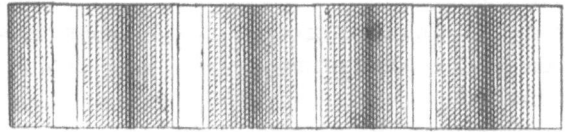

Fig. 69. Position of dark and bright bands in a quartz wedge in monochromatic light, when the wedge is in a diagonal position between crossed Nicol prisms.

wedge may be used to determine wavelengths. Let us consider this problem in more detail.

It is clear from Fig. 70 that the distance between neighboring black bands in the wedge is given by

$$a = \frac{d_2 - d_1}{\operatorname{tg} \alpha}, \tag{85}$$

where d_1 and d_2 are the wedge thicknesses corresponding to the two black bands.

Using formula (64) which was derived earlier, we have

$$\Gamma = d\,|\,n' - n''\,|$$

and taking into account the fact that the path difference Γ increases in the transition from d_1 to d_2 by one wavelength, we may write

$$\Gamma_2 = d_2\,(n_e - n_o),$$
$$\Gamma_1 = d_1\,(n_e - n_o).$$

Hence

$$\Gamma_2 - \Gamma_1 = \lambda = (n_e - n_o)\,(d_2 - d_1),$$
$$d_2 - d_1 = \frac{\lambda}{n_e - n_o}.$$

Substituting this value of $d_2 - d_1$ into (85), we obtain

$$\lambda = a\,(n_e - n_o)\operatorname{tg} \alpha.$$

Since the birefringent power $n_e - n_o$ of quartz is ·practically independent of λ (it is equal to 0.0090 for red and 0.0095 for violet rays) it follows that, very approximately, the wavelength is given by the expression

81

TABLE IV

Interference Colors in Crystal Plates as a Function of the Path Difference

Order of color	Path difference in mμ	Color for crossed Nicol prisms	Color for parallel Nicol prisms
1	0	black	white
	100	grey	bright yellow
	260	white	red
	300	yellow	violet
	450	brown	light blue
	500	orange	
	550	red I	bright green
2	575	violet	yellow-green
	590	indigo	yellow
	700	light blue	orange
	800	green	red
	850	yellow-green	violet
	910	yellow	indigo
	950	orange	light blue
	1100	red II	green
3	1130	violet	yellow-green
	1150	indigo	yellow
	1330	aquamarine	red
	1430	yellow-green	violet
	1500	meat-red	aquamarine
	1530	red III	green
	1650	bright violet	bright yellow-green
4	1710	bright green	rose
	2000	bright grey	bright grey
	2050	rose	bright red

$$\lambda = 0,009\, a\, \mathrm{tg}\, \alpha. \qquad (86)$$

which may be used to determine wavelengths by means of a quartz wedge.

Quartz Wedge in White Light. So far we have assumed that the quartz wedge is placed between crossed Nicol prisms and is observed in monochromatic light. If instead of monochromatic light one takes white light, then at points which correspond to different wedge thickness, the wedge will have different hues. We know already how to determine these hues approximately. Figure 68 may be used for this purpose in the present case. Starting with a given thick-

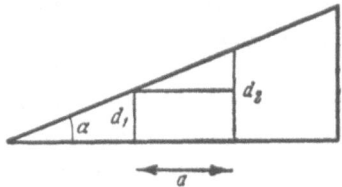

Fig. 70. Drawing illustrating the derivation of formula (85).

ness d, we draw through this point a vertical line. It intersects the curve at points which correspond to the different values of J_λ for different wavelengths. If we add all these values together, in accordance with the color mixing rules given above, we obtain information on the hue of the wedge at the given point.

Table IV gives the observed sequence of colors which is obtained with crossed and parallel Nicol prisms.

In this table, the color of the crystal plate is given not as a function of the thickness d, but as a function of the path difference Γ, which is proportional to the plate thickness. If the interference hue of any crystal plate, not necessarily of quartz, can in general be characterized by some quantity, then the latter can only be the path difference. The transition from the path difference to the thickness d is very simple in any given case because according to (64), the relation between them is

$$\Gamma = d\,|n' - n''|.$$

Let us consider the variation in the interference hue of a crystal plate placed between crossed Nicol prisms. It follows from the above table that crystal plates have no hue for small path differences. A very thin crystal plate placed between crossed Nicol prisms does not produce any noticeable optical effects and will appear dark since crossed Nicol prisms do not transmit light. As the path difference is slowly increased, the system will begin to transmit light and the plate will pass through various shades of grey colors. When the path differ-

83

ence is about 260 mμ, the plate will become white (lower order white light). Next, beginning with yellow light, the hues will follow in the order in which they are defined as increasing in the color rosette. When the path difference is considerably increased, the colors become less and less saturated passing through rose and greenish shades into higher-order white color.

The variation in the interference color as the thickness of the plate is increased, is clearly an aperiodic phenomenon since none of the colors are quite identical. Nevertheless, one can speak about some periodicity. This refers above all to red color, which repeats itself in various shades several times, and gives a basis for dividing interference colors into orders which correspond to increases in the path difference of 550 mμ. Such a path difference is equal to the wave length of bright yellow-green rays. The red-violet shades which for crossed Nicol prisms divide the interference hues into orders, are complementary to these rays. The human eye can distinguish about six orders. The most saturated are the second and third order colors. One notes the very rapid change in color in the boundary between the first and second orders. In this region, the violet band is the narrowest and this color has received the name of violet sensitive tint. With an extremely small change in the path difference this tint changes to blue or red. Crystal plates (quartz or gypsum) which, when placed between crossed Nicol prisms give this violet sensitive tint, are used to determine very small path differences. It is sufficient to place the thinest piece of mica on such a plate in order to change the violet hue quite noticeably and either increase or decrease it.

The peculiar repeatability of shades in the aperiodic alternation of interference colors is a kind of beat phenomenon (p. 12) which appears whenever two or a number of periodic phenomena having different periods are superimposed on each other. The alternation of colors in the quartz wedge which is observed with white light, may be looked upon as a result of the superposition of a number of interference patterns obtained in monochromatic light, each of which has the form of a sequence of dark and bright bands, but in different patterns the bands are at a different distance from each other. Some idea as to the character of such beats is given by Fig. 71 (opposite p. 90), in which three sets of bands of different color and different period are superimposed on each other.

So far we have been concerned with interference colors of a quartz wedge placed between crossed Nicol prisms. As we know, colors observed with parallel Nicol prisms should be strictly complementary, according to their spectral composition, to the corresponding colors with crossed Nicol prisms. Bearing this in mind, it is easy to understand why, in the case of a very thin plate and parallel Nicol prisms, white color is complementary to black, and lower order white color is complementary to red for a path difference of 260 mμ and crossed Nicol prisms, while in the case of higher order white color, the complementary color is also white.

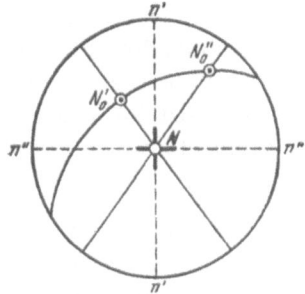

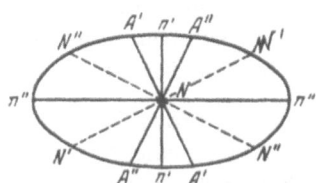

Fig. 72. Drawing illustrating the derivation of Fresnel's theorem. N_0', N_0'' are the exits of the optic axes, N the exit of the normal to the plate, $n'n'$, $n''n''$ are the vibration directions in the plate, the arc $n'Nn'$ bisects the angle $N_0'NN_0''$.

Fig. 73. Drawing illustrating the derivation of Fresnel's theorem. Section of the indicatrix cut by the plane of the crystal plate; $A'A'$, $A''A''$ are the diameters of circular sections and $N'N'$, $N''N''$ are the normals to these diameters.

Dependence of Transmission on the Quantity $(n' - n'')$. In general, this dependence has the same character as the dependence of the transmission on the magnitude of d, since both these quantities enter into the transmission formula in a similar way.

In crystal-optical measurements, the quantity $(n' - n'')$ represents the difference in the refractive index of two monochromatic rays of given color, leaving the crystal plate in the same direction, namely, the direction of the axis of the polarizing instrument. In general, this quantity depends on the inclination of the plate to the axis of the instrument, but it is difficult to determine the effect of $(n' - n'')$ alone on the transmission since, when the plate is inclined to the axis of the instrument, the quantities α, β, d may also change. Below, we shall discuss special cases of the inclination of the crystal plate, which are the most important in practice, but first we must give an account of the following theorem due to Fresnel.

85

Fresnel's Theorem. This theorem may be formulated in the following way: the vibration directions n' and n'', corresponding to an arbitrary wave normal N, inside the crystal, bisect the angle formed by the planes NN_o'' and NN_o', which contain the normal N and the optic axes N_o' and N_o'' of the crystal.

Let the wave normal N be perpendicular to the plane of the drawing (Fig. 72) so that the vibration directions n', n'' lie in the plane of the drawing. The optic axes N_o' and N_o'' will then be at an angle to N. Accordingly, in the stereographic projection shown in the drawing, the exits of the optic axes are inside the projection circle. Fresnel's theorem states that in order to find the vibration directions which in the drawing are shown by dotted lines (and the heavy cross in the center), one must draw in the planes NN_o' and NN_o'' and bisect the angles between them. The corresponding bisectors then give the required vibration directions.

In order to prove Fresnel's theorem, let us take one of the central sections of the indicatrix (Fig. 73). In general, this section is an ellipse whose principal axes coincide with the vibration directions corresponding to the normal N to the plane of the ellipse. This plane (if it does not accidentally pass through the mean axis of the indicatrix) must intersect the two circular sections along their diameters $A'A'$ and $A''A''$. Since these diameters are, of course, equal and are at the same time diameters of the ellipse, they must be located symmetrically with respect to n' and n''. The normals N' and N'' to these diameters will necessarily be also symmetrical relative to n' and n''. Since $A'A'$ is one of the diameters of the circular section of the indicatrix, it follows that the optic axis N_o' should be at right angles to $A'A'$, i.e., it should lie in the plane $N'N'$, perpendicular to the plane of the drawing. Similarly, the second optic axis N_o'' should lie in the plane $N''N''$, perpendicular to the plane of the drawing. In other words, the plane $N'N'$ coincides with the plane NN_o', which contains the normal N and the optic axis N_o', while the plane $N''N''$ coincides with NN_o''. Since n' and n'' bisect the angles between these planes, the theorem is proved.

Inclination of a Uniaxial Crystal Plate. Let us consider two simple, but also the most important, special cases of inclination of the plate.

1. The uniaxial crystal plate is placed between crossed Nicol prisms in the extinguished position, and is rotated at an angle to one of its vibration directions. We shall not consider all the possible inclinations of the plate and, using the stereographic projection, will

assume that in the original position the plate lies in the plane of the drawing and a parallel beam of light is incident normally on it in the direction toward the observer. In this case, the vibration directions n', n'' in the plate in its original position and the vibration directions in the Nicol prisms I, II will coincide (Fig. 74). The exit of the optic axis is shown in the drawing by the point n_e, and the circular section of the indicatrix by the arc ABA.

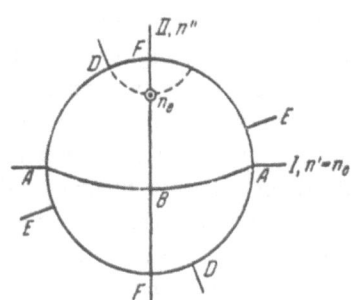

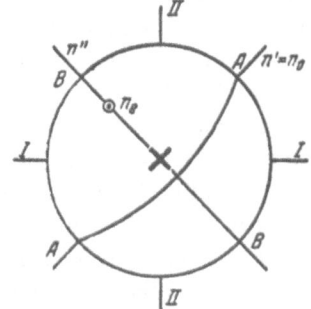

Fig. 74. Drawing illustrating the effect of the tilting of a uniaxial crystal plate from the extinction position in the general case. When the plate is rotated about AA the extinction is preserved. When the rotation is about FF the extinction is lost.

Fig. 75. Drawing illustrating the effect of the tilting of a uniaxial crystal plate from the diagonal position between crossed Nicol prisms. When the plate is rotated at an angle to AA extinction of the plate is possible; it will be preserved when the plate is then rotated about the vertical axis. When the plate is rotated at angle to BB extinction is again possible but it will not be preserved when the plate is then rotated about a normal to it.

If the plate is rotated about AA, the optic axis n_e will move in the plane $n_e B$ and the direction of the projection of the optic axis will remain constant. Since this direction is also the vibration direction in the plate, the plate will remain extinguished.

If the plate is rotated about the line FF, then the end point of the optic axis will describe a small circular arc $n_e D$ and will finally lie in the plane of the drawing on the line DD. Since the new vibration directions in the plate (DD and EE) are now no longer coincident with the vibration directions in the Nicol prisms, the above rotation of the plate will bring it out of the extinguished position.

Thus, in the case of uniaxial crystals under above conditions, one of the extinguished positions is preserved and the other is not.

So far we have considered the general case in which the plate intersects the indicatrix at an oblique angle with respect to the optic axis. If the plate is kept parallel to the optic axis, then one can easily imagine that both the extinguished positions are preserved. When the plate is cut perpendicular to the optic axis, it will remain extinguished not only when it is rotated at an angle to AA and FF, but also when it is rotated about its normal.

2. A uniaxial crystal plate is placed between crossed Nicol prisms in a diagonal position and is rotated at an angle to one of its vibration directions (Fig. 75). If the plate is rotated at an angle to AA, then it will remain in the diagonal position since the optic axis whose exit is at n_e will remain in the plane $n_e B$. At the same time, the refractive index n'' can either increase or decrease, but must lie between $n'' = n_0$ and $n'' = n_e$, while d (the path length along the wave normal) will always increase. Thus, in this case the inclination of the plate may lead to either an increase or a decrease in the hue. This also follows from the transmission formula which now assumes the form

$$ J = \sin^2 \frac{\pi d \, | \, n' - n'' \, |}{\lambda}, $$

where the variables are now d and $(n' - n'')$.

It is important to note that if the plate can be inclined so that the incident rays pass through the plate along its optic axis, and hence will leave it along the axis of the instrument, then the plate will be extinguished for any original d.

If the plate is rotated at an angle to BB, then the point of exit of the optic axis will describe on the sphere a small circle and consequently, the plate will leave the diagonal position (α and β will change), and since both n' and n'' and also d vary at the same time, the plate will in general change its hue.

Inclination of a Biaxial Crystal Plate. Let us consider two important special cases of inclination of a biaxial crystal plate.

1. A biaxial crystal plate is placed between crossed Nicol prisms in the extinguished position and is rotated at an angle to one of its vibration directions. Figure 76 shows the original position of the plate in the general case, i.e., for a plate cut in an arbitrary manner. Using Fresnel's construction, it is easy to show (cf. Fig. 72) that in this case, when the plate is rotated about the line AA the optic axial plane of the crystal will change its orientation in space and, as a result, the plate should leave the extinguished position. This should also occur

when the plate is rotated about the line *BB*. It follows that in biaxial crystals under the above conditions, both the extinguished positions are lost.

Finally, if the plate is cut along one of the symmetry planes of the indicatrix, both the extinguished positions will be preserved under the above rotations.

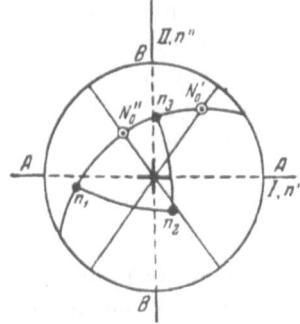

Fig. 76. Drawing illustrating the effect of the tilting of a biaxial crystal plate from the extinction position. General case; extinction is lost when the plate is rotated both about *AA* and *BB*.

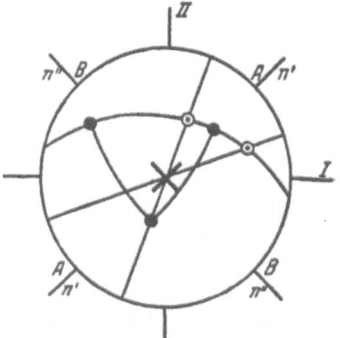

Fig. 77. Drawing illustrating the effect of the tilting of a biaxial crystal plate from the diagonal position between crossed Nicol prisms. General case; the hue changes; extinction is possible when the plate is rotated about *AA* or *BB*.

2. A uniaxial crystal plate is placed between crossed Nicol prisms in a diagonal position and is rotated at an angle to one of its vibration directions.

Figure 77 shows the original position of the plate cut in an arbitrary way. It is easy to see that in this case, when the plate is inclined, all the three parameters d, n', n'', in the transmission formula (73) should change and this should lead to a change in the hue of the plate.

The special case where in the original position of the plate, the optic axial plane is perpendicular to it (Fig. 78), is particularly important. In this case (the angle between the acute bisectrix and the normal to the plate is usually small), rotation about the line AA should give two extinguished positions. These positions will be preserved if the plate is subsequently rotated about the axis of the polarizing instrument.

Concluding Remarks on the Transmission Formula. Let us return to the transmission formula for a crystal plate placed in the diagonal position between crossed Nicol prisms (83), and rewrite this formula in the following form

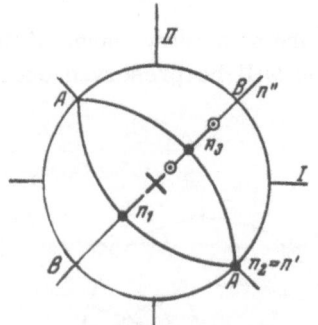

Fig. 78. Drawing illustrating the effect of the tilting of a biaxial crystal plate from the diagonal position between crossed Nicol prisms in the special case where the optic axial plane is perpendicular to the plate. When the plate is rotated about AA extinction is possible twice. When it is rotated about BB the interference hue changes; extinction is also possible.

$$J = \sin^2 \frac{\Delta}{2} = \sin^2 \frac{\pi \Gamma}{\lambda}.$$

It is clear from this formula that the transmission will be zero when $\Gamma = n\lambda$, where n is an integer. This result is in an apparent contradiction to the well-known fact that, usually total extinction of the interfering rays is observed when $\Gamma = (2n + 1)\frac{\lambda}{2}$ and not when $\Gamma = n\lambda$. This is explained by saying that in the two cases, the path difference does not mean the same thing. In the transmission formula one is concerned with the path difference between two waves after they have left the crystal plate and before they enter the second Nicol prism. This difference is the final path difference only when the second Nicol prism is parallel to the first. If, however, the Nicol prisms are crossed, an extra path difference of $\frac{\lambda}{2}$ due to the second Nicol prism must be added. One therefore speaks of the loss (or gain) of one half-wavelength when the interference takes place with crossed Nicol prisms. It is important to note that when white light is used, this half-wavelength loss is experienced by each pair of interfering monochromatic waves which are part of the white light, and hence waves with different λ have identical G but different Γ (cf. (65)). A change in the thickness of the plate cannot lead to this effect.

Our second remark must explain why both Nicol prisms are necessary in the observation of interference in crystal plates. We know that in the case of thin transparent plates, the necessary and sufficient condition for the interference to take place is the existence of a path difference Γ. In crystal plates this path difference depends only on d and $(n' - n'')$:

$$\Gamma = d \, | \, n' - n'' \, |.$$

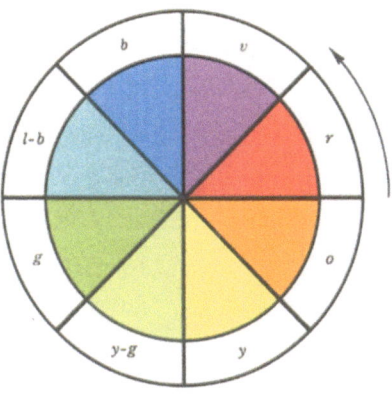

Fig. 60. An eight-step color rosette. The arrow shows the
order of increasing hue.

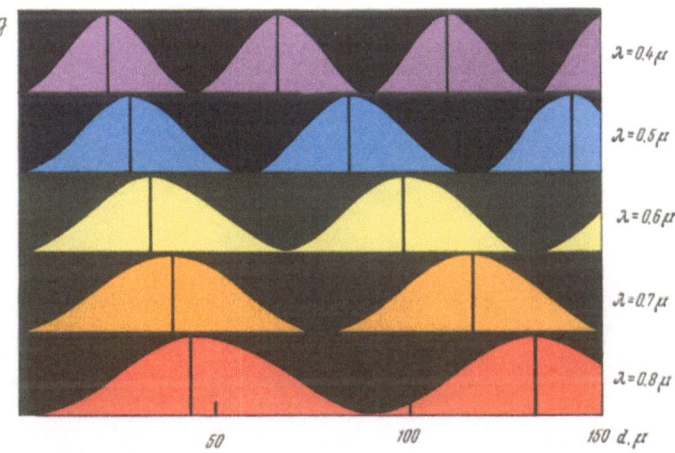

Fig. 68. Dependence of the transmission J on the thickness d (in microns)
of a quartz wedge for different wavelengths, the wedge being placed in a
diagonal position between crossed Nicol prisms. The colors shown do not
reproduce the colors actually observed very well.

Fig. 71. Parallel superposition of three sets
of bands of different color and period.

It would therefore appear that interference phenomena should also take place in the absence of Nicol prisms. These phenomena do, of course, take place, but they cannot be observed directly with the eye for the following two reasons.

1. The vibration directions of the two monochromatic waves which appear in the crystal as a result of double refraction are perpendicular to each other. The superposition of these waves in general leads to elliptical vibrations (p. 145) and hence they cannot completely extinguish each other. A second Nicol prism is necessary in order to reduce the vibrations in the two waves into a single plane, and only under these conditions can the interference be seen directly by the eye, by observing changes in the intensity of the transmitted light.

2. In order to understand why the first Nicol prism is necessary, we recall that natural light consists of quanta polarized in different planes. This means that the angle α in the transmission formula is constantly and randomly changing between 0 and π with an enormous mean frequency. The overall effect of such a change in α will, as far as the eye is concerned, be equivalent to a rapid rotation of the first Nicol prism. However, we know that under these conditions the interference hue should be equally rapidly and periodically replaced by the complementary hue, and since the eye cannot follow this constant interchange of hues, the observed hue will be white if white light is used.

Dispersion of Vibration Directions. So far, in using the transmission formula for white light, we assumed (without especially bringing the reader's attention to the fact) that the vibrations in waves of different frequency in a crystal plate take place in two, and only two, directions, namely, n' and n''. In fact, this is not always the case. The noncoincidence of vibration directions for waves of different frequency is known as the dispersion of vibration directions. It is a simple consequence of indicatrix dispersion which was discussed earlier. It is easy to verify, using the Fresnel construction, that the dispersion of vibration directions does not occur in uniaxial crystals, and cannot be observed in biaxial crystals in sections perpendicular to the symmetry planes of the indicatrix family. Let us consider the extinction of a crystal plate in which dispersion of vibration directions is possible, when the plate is rotated about its normal, and the observations are carried out in white light and with crossed Nicol prisms (Fig. 79).

The presence of the above dispersion means that the vibration directions of waves of different color lie within an angle $\Delta\alpha = \alpha_v - \alpha_r$, (which is usually very small) where α_v is the angle between the vibration directions in the first Nicol prism and the vibration directions in the extreme violet ray, and α_V has a similar meaning for the extreme red ray. If the plate is rotated in, say, a counterclockwise direction, not all the waves will be extinguished at the same time, but waves of a given color will be successively extinguished in the same order in which they occur in the spectrum. Since $\Delta\alpha$ is usually

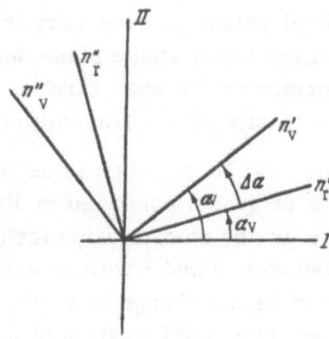

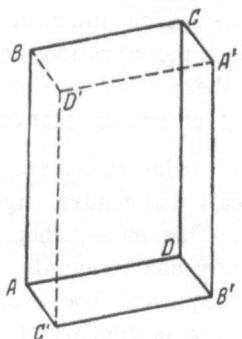

Fig. 79. Drawing illustrating the effect of dispersion of vibration directions. The vibration directions for violet and red rays are not the same. The sign of the dispersion changes when the plate is turned over.

Fig. 80. Drawing showing the possibility of distinguishing between the faces $ABCD$ and $A'B'C'D'$ an oblique parallelepiped.

small, it follows that when the waves of one color are fully extinguished, the intensity of the others will be negligible, and hence as the plate is rotated, covering the whole interval $\Delta\alpha$, the observer will see (under conditions indicated in the diagram) a change from brown (less the cold shades) to a dull blue hue. When the plate is rotated in the opposite direction, the alternation of these dark colors will also take place but in the opposite direction. This will be the way in which the plate will be "extinguished" in the presence of dispersion of vibration directions. In the absence of this form of dispersion, no difference should be observed in the process of extinction when the plate is rotated in a clockwise or counterclockwise direction.

The phenomenon of dispersion of vibration directions is of major theoretical interest since it may be used to distinguish between the two sides of the plate according to the sign of the rotation, i.e., the right side from the left, where the right side is defined by $\alpha_V > \alpha_r$ and the left by $\alpha_V < \alpha_r$.

In this connection we must make the following digression. Consider an oblique parallelepiped of the general form shown in Fig. 80. The question arises whether it is possible to distinguish the face $ABCD$ in some way from the parallel face $A'B'C'D'$. This question has almost always a negative answer since the two faces are equal to each other. However, this does not take into account the fact that the equality is not combinable but a mirror one, since the first face can only be superposed on the second by an inversion operation (reflection at the center of symmetry) and inversion itself, similarly to a reflection in a plane, is an operation which transforms left into right. It follows that the faces $ABCD$, and $A'B'C'D'$ can only differ in respect of left-handedness or right-handedness, i.e., the sign of rotation. In fact, in the case of the $ABCD$ face, the sequence of corners denoted by these letters, is a right-handed one while in the case of the $A'B'C'D'$, face, the sequence of similar corners is left-handed provided it is looked at under similar conditions, i.e., not through the $ABCD$ face.

In optics, one often uses the so-called Fermat's rule, according to which the direct path of a ray differs from the reverse path only by the sign of incidence. The above example shows, however, that in the case of transmission of light through a crystal plate in the direct and reverse directions, there is also a change in the sign of rotation (cf. pp. 19, 115 and 116).

Compensation of Double Refraction. If one takes two identical crystal plates and then places one on top of the other in such a way that the n' and n'' directions in the one plate coincide with the similar directions in the other plate, then as far as optical phenomena are concerned, this double plate behaves as a single plate of double thickness. If this combination of plates is placed between Nicol prisms, one observes a path difference which is twice that observed in each of the component plates taken separately.

Let us now cross the plates, i.e., place them so that the n' direction in one of the plates coincides with the n'' direction in the other, and conversely, the n'' direction in the one with the n' direction in the other. In this case waves propagated in the first plate with the lower (higher) velocity, will be propagated in the second with the higher (lower) velocity. As a result, the path difference Γ_1, in the first plate will be fully compensated by the path difference Γ_2 in the second plate. With crossed Nicol prisms we will therefore see a dark field.

When the thicknesses are unequal $(d_1 > d_2)$, the compensation will not be complete and

$$\Gamma_1 - \Gamma_2 = (d_1 - d_2)\,|\,n' - n''\,|.$$

If the two plates are made of different crystals, or are cut from a single crystal but in different ways, the birefringent powers $|n_1' - n_1''|$ and $|n_2' - n_2''|$ will not, in general, be equal. In this case, total compensation of double refraction can only be obtained with unequally thick plates provided the condition

$$d_1\,|\,n_1' - n_1''\,| = d_2\,|\,n_2' - n_2''\,|$$

is satisfied. These thicknesses should clearly be inversely proportional to the corresponding birefringent powers, i.e.,

$$\frac{d_1}{d_2} = \frac{n_2' - n_2''}{n_1' - n_1''}. \tag{87}$$

So far, we have considered the compensation of double refraction of monochromatic waves. When the observations are carried out in white light, total compensation is only possible for two identical plates. In practice, one often uses (in the determination of an unknown birefringent power from a known one) incomplete compensation of double refraction of the two brightest yellow spectral lines, and ignoring the dispersion of vibration directions.

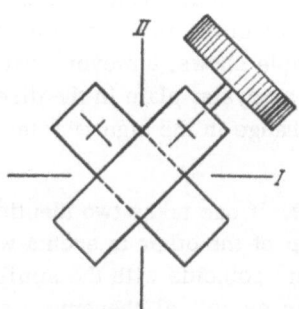

Fig. 81. A compensator with a tilted crystal plate.

Compensators. Various types of compensators are used in double refraction studies. The simplest of these is the quartz wedge which was described earlier. In the compensation of double refraction with the aid of a quartz wedge, the crystal plate under investigation is placed on the stage of a polarizing microscope in a diagonal position and between crossed Nicol prisms. The thin edge of the wedge is introduced above the plate, which is also in the diagonal position but is crossed with the quartz wedge, until maximum extinction is obtained. When the plate is removed it is possible to determine the path difference produced in the plate from the color of the wedge at the point where the maximum extinction was observed. Table IV, which was given earlier, (or more detailed tables taken from special texts) may then be used in this determination. By dividing the value of Γ obtained in this way by the thickness d of the plate, we find its birefringence.

The Babinet compensator is very widely used. This compensator consists of a thin calcite plate, cut perpendicular to the optic axis. The plate can be rotated about a horizontal axis lying in its plane. It is placed above the crystal plate under investigation, and is always in a diagonal position with respect to the vibration directions in the crossed Nicol prisms. By rotating the calcite plate it is possible to extinguish the crystal plate under investigation. The corresponding angle of rotation, which can be read off a micrometer scale, can be converted into the required path difference with the help of tables appropriate to the given compensator (Fig. 81).

The Babinet compensator is one of the earliest compensators and is usually described in physics textbooks.

Interference in Convergent Light. So far we have considered interference phenomena in crystal plates when a parallel beam of polarized light is passed through them. They include phenomena which are observed when the crystal plates are rotated about horizontal and vertical axes. The use of convergent light enables one to observe, in a single interference pattern, all those phenomena which in the case of parallel light can only be observed successively by rotating the plates about the above axes.

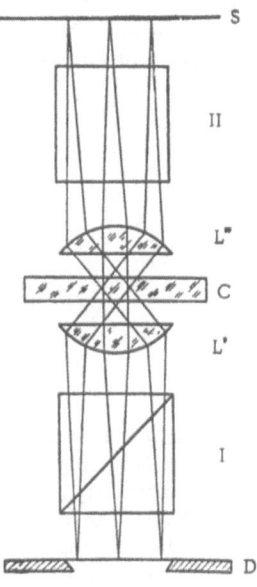

Fig. 82. Observation of interference patterns in convergent light. Light from the luminous plane D passes through the Nicol prism I, the crystal plate C placed between the lenses L', L*, and the Nicol prism II. The interference pattern is obtained at S. The plane D lies in the principal focal plane of the lens L'; the plane S lies in the principal focal plane of L*.

The conditions under which interference patterns can be obtained in convergent light are shown schematically in Fig. 82. The crystal plate is placed between two lenses and the two Nicol prisms. A luminous plane (not a point) placed in the focal plane of the first lens is the source of light giving rise to the interference pattern. The luminous plane can be in the form of a white sheet of paper, matt glass, or a suitably illuminated diaphragm. Such a source ensures that within the given angle of convergence, there will be, inside the crystal, a sufficiently wide beam of parallel rays (and not just one ray as was the case with a point source of light) which will be focused at a point with sufficient brightness by the second lens. The aggregate of such points gives the total interference pattern which, similarly to any real image, can be observed by the eye from different viewing positions. The form of these interference patterns depends on certain special surfaces and families of curves which we shall now consider.

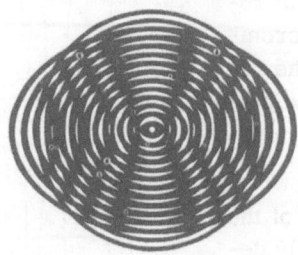

Fig. 83. Formation of surfaces of equal path difference in a uniaxial negative crystal. The diagram shows intersecting concentric circular and elliptical waves. Each circle is inscribed in the corresponding ellipse.

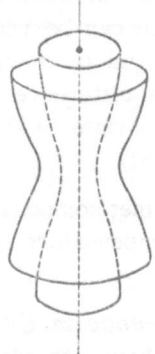

Fig. 84. Surface of equal path difference in uniaxial crystals.

Surfaces of Equal Path Difference. Let us imagine a point source of monochromatic light placed within a uniaxial crystal which, in accordance with our earlier discussion, produces spherical and ellipsoidal waves within the crystal. Figure 83 shows an instantaneous "photograph" of the wave field inside the crystal in a plane section along the optic axis. It is clear from this figure that, under the above conditions, stationary group waves should be formed, which in the drawing pass through the points of intersection of the black circles and ellipses, i.e., through points of equal ray path difference Γ_S (so far we have only dealt with the wave path difference Γ_N, which we denoted by Γ without the subscript N; the two path differences are related by the equation $\Gamma_N = \Gamma_S \cos \alpha$, where α is the angle between a ray and the wave normal). It is not difficult to verify by a simple calculation that the group surface nearest to the source of light corresponds to a path difference of one wavelength λ_S, the next surface corresponds to points with a path difference of two wavelengths $2\lambda_S$, and so on.

Some idea about the form of the family of such surfaces for uniaxial crystals can be obtained from Fig. 84, which shows two such surfaces nearest to the optic axis. Each of the surfaces is a surface of revolution and hence a section perpendicular to the optic axis gives a set of concentric circles.

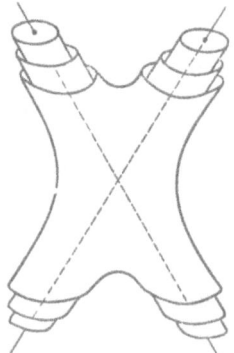

Fig. 85. Formation of surfaces of equal path difference in biaxial crystals. The diagram is based on ellipses intersecting circles (cf. Fig. 39).

Fig. 86. Surfaces of equal path difference for biaxial crystals.

Surfaces of equal path difference for biaxial crystals can be constructed in a similar way (Figs. 85 and 86). A characteristic property of a family of such surfaces is the fact that they include a pair of straight lines which represent zero path difference. Sections perpendicular to the acute bisectrix give curves similar in form to the Cassini ovals. As they approach the biradials, they degenerate into a curve similar to a lemniscate (a figure of eight) and then rings of irregular form (Fig. 87). Our discussion of surfaces of equal path difference (surface of equal Γ_S)

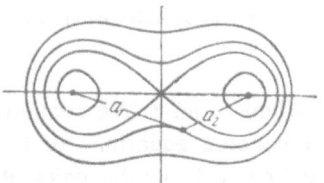

Fig. 87. Section through a family of Bertin's surfaces.

was based on ray velocity surfaces (v_S). If this derivation is based on the normal velocity surface (v_N) one obtains a surface of equal Γ_N. Such a surface is known as a Bertin surface. Bertin surfaces do not differ very much from the Γ_S surfaces.

97

Since the above surfaces will only be used in qualitative discussions of interference patterns, we shall not distinguish between surfaces of equal Γ_S and equal Γ_N and shall call them both isochromatic surfaces.

Equations of Isochromatic Surfaces and Isochromes. For a given path difference Γ, the isochromatic surface of a uniaxial crystal is approximately given by

$$r = \frac{\Gamma}{\sin^2 \varphi},$$ (88)

where r is a variable radius vector and φ is the angle between this radius vector and the optic axis of the crystal. In the case of biaxial crystals, the approximate equation of an isochromatic surface may be written down in the form

$$r = \frac{\Gamma}{\sin \varphi_1 \sin \varphi_2},$$ (89)

where φ_1, φ_2 are the angles between the variable radius vector and the optic axes of the crystal (Fig. 88).

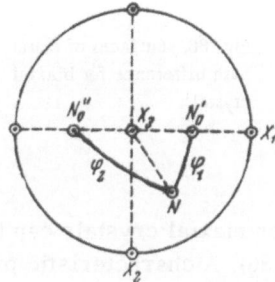

Fig. 88. Drawing illustrating the meaning of the quantities φ_1 and φ_2 which enter into equation (89).

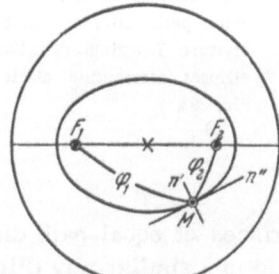

Fig. 89. Drawing illustrating the concept of a spherical ellipse.

When r and Γ are constant, the equations of isochromatic surfaces (88) become equations of isochromatic curves inscribed on a sphere (isochromes). In the case of uniaxial crystals, the isochromes form a family of circles on the sphere, with a common "center". In the case of biaxial crystals equation (89) becomes the equation of a family of isochromes:

$$\sin \varphi_1 \cdot \sin \varphi_2 = \text{const},$$ (90)

which resembles the equation describing a family of Cassini ovals

$$a_1 \cdot a_2 = r^2 = \text{const},$$ (91)

where a_1, a_2 are the distances from any point on the Cassini curve from its foci (Fig. 87). If the distance between the foci is denoted by $2d$ and $\frac{d}{r} = \varepsilon$, then for $\varepsilon = 0$ the Cassini ovals become circles, when $\sqrt{0.5} > \varepsilon > 0$ the curves resemble ellipses, and when $\sqrt{0.5} < \varepsilon < 0$ they resemble ovals with two indentations; when $\varepsilon = 1$ the curves are lemniscates resembling a figure of eight, and when $\varepsilon > 1$ the curve divides into a pair of irregular rings.

The isochromes inscribed on a sphere have qualitatively similar properties to Cassini ovals. When φ_1, φ_2 are small, equation (90) in general goes over to equation (91).

Isogyres. Consider two points F_1, F_2 (exits of the optic axes) on a spherical surface and let the point M be at angular distances φ_1, φ_2 from them (Fig. 89). If the point M is displaced in such a way that

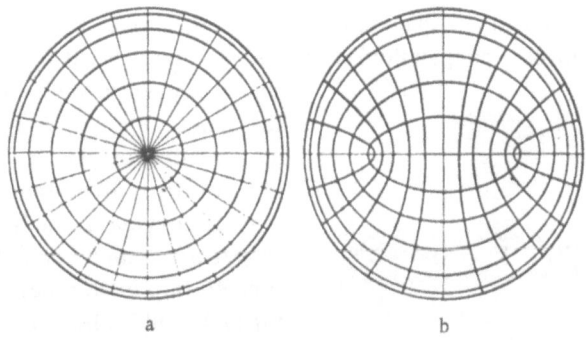

Fig. 90. Coordinate nets on a sphere used to determine vibration directions. a) Uniaxial crystals(a set of meridians and parallels); b) biaxial crystals (a set of spherical ellipses).

$\varphi_1 + \varphi_2$ remains constant, it will describe a curve which may be called a spherical ellipse with foci at F_1 and F_2. Such a curve has a number of properties similar to the properties of an ordinary plane ellipse, in particular, the normal n' to the spherical ellipse which passes through the point M is a bisector of the angle formed by the arcs φ_1, φ_2. According to the Fresnel theorem which was established earlier, the normal n' and the tangent n'' to the spherical ellipse coincide in direction with the vibration directions referred to the wave normal passing through the point M.

Since the optic axes cut the spherical surface at four points, any pair of which (belonging to different axes) can be taken as the foci of the ellipses, it follows that altogether only two systems of orthogonal ellipses can be drawn on a sphere. An elliptical net drawn on a sphere or a plane in stereographic projection (Fig. 90 b) may therefore be used to determine the vibration directions of wave normals. In the case of uniaxial crystals, the points F_1, F_2 collapse into a single point on the optic axis. The elliptical net therefore becomes an ordinary set of circles, i.e., the meridians and parallels of Fig. 90 a.

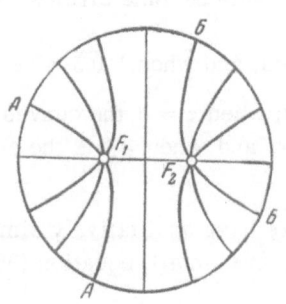

Fig. 91. Isogyres constructed from the set of ellipses in the previous drawing.

Using the above net of ellipses (circles) on a sphere, one can construct the corresponding net of isogyres which are curves drawn on a sphere which pass through points with the same vibration directions (Fig. 91). Since in waves propagated along the optic axis, the vibrations are in all directions perpendicular to the wave normal, it follows that all the isogyres should intersect at the points F_1, F_2. A pair of isogyres corresponds to each pair of vibration directions n', n''. In biaxial crystals such a pair of isogyres, which can be transformed into each other by rotation through 180° about the acute bisector, for example, the isogyres AA and BB form such a pair. As the foci F_1, F_2 approach each other, a biaxial crystal gradually becomes uniaxial. The system of isogyres then gradually becomes the usual system of meridians. Thus, in uniaxial crystals, to each separate isogyre there corresponds a spherical "biangle" formed by two semimeridians intersecting each other at right angles, while conjugate isogyres are any two "biangles" of the aggregate, which form a pair of meridians intersecting each other at right angles.

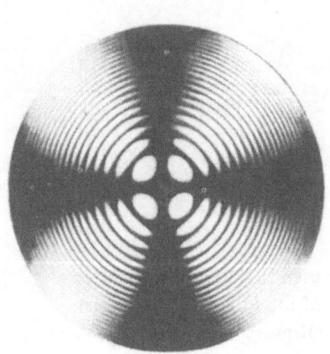

Fig. 92. Interference pattern for a uniaxial nonrotating (see later, pp. 111 and 142) crystal in convergent light cut perpendicular to the optic axis and placed between crossed Nicol prisms. Black cross and colored rings observed in white light. Black cross with dark and bright, equally colored rings, observed in monochromatic light.

Interference Patterns Obtained with Uniaxial Crystals in Convergent Light. If a uniaxial crystal plate cut perpendicular to the optic axis is observed in monochromatic light between crossed Nicol prisms, the corresponding interference pattern has the form of concentric light and dark rings intersected by a dark cross concentric with them, whose arms are parallel to the vibration directions in the Nicol prisms (Fig. 92). The bright and dark curves are circular because under these conditions, rays which are equally inclined to the optic axis, i.e., those having the same d and ($n'-n''$), have the same path difference. The center of the field of view is dark because the final path difference at this point is $1/2 \lambda$. It consists of zero path difference in the plate and $1/2 \lambda$ path difference due to crossed Nicol prisms (p. 90). The first bright ring corresponds to a final path difference of λ, which consists of a path difference of $1/2 \lambda$ in the plate and an equal path difference due to the crossed Nicol prisms. The first dark ring corresponds to a path difference of $\lambda + 1/2 \lambda = 3/2 \lambda$, and so on. Points lying on the diagonals of the dark cross have the highest intensity. This is explained by the fact that the vibration directions n', n'' in the corresponding rays are diagonal relative to the vibration directions in the Nicol prisms. The presence of the dark cross is due to the fact that the vibration directions in the corresponding rays are in the vibration plane of one of the Nicol prisms. Such rays are not doubly refracted in the crystal plate and are not transmitted by the second Nicol prisms. The larger the wavelength of the monochromatic light, the larger are the circles in the interference pattern, since in order to obtain the same wave path difference (65)

$$G = \frac{\Gamma}{\lambda}$$

the path difference must increase in the same proportion as λ (i.e., the angle between the ray and the optic axis must increase).

If the observations are carried out in white light, the rings assume a spectrum of hues. Since the path difference increases with the inclination of the rays to the optic axis, the sequence of colors in the rings along a radius will be the same as in a quartz wedge as its thickness increases.

The change in the hues when the crossed Nicols are replaced by parallel Nicols is the same as before, namely, the interference pattern with parallel Nicol prisms will be strictly complementary (according to the spectrum) to the pattern obtained with crossed Nicol prisms. The bright monochromatic rings are replaced by dark rings and, vice versa, rings having mixed colors are replaced by the complementary mixed colors, the dark cross is replaced by a white cross if white light is used, and by a monochromatic cross when monochromatic light is used.

If the plate is cut accurately perpendicular to the optic axes, the interference pattern will remain unaltered when the plate is rotated

Fig. 93. Schematic drawing of an interference pattern obtained with a uniaxial crystal in convergent light when the crystal is cut obliquely relative to the optic axis and is placed between crossed Nicol prisms. When the crystal plate is rotated about one of its normals, the arms of the cross move parallel to themselves.

or displaced in its own plane. If the plate is cut obliquely, the center of the cross will not coincide with the center of the field of view and can even lie outside this field (Fig. 93). When the plate is rotated about a normal to it, the center of the dark cross will describe a circle in the field of view but the arms of the cross will be displaced parallel to themselves. If the center of the cross cannot be seen, then the arms of the cross will alternately enter the field. The parallel displacement of dark arms of the cross (when the angle between the plane of the plate and the optic axis is not too large) may be used to distinguish uniaxial from biaxial crystals, since in the latter crystals the arms of the cross will bend as the plate is rotated.

In plates cut parallel to the optic axis, and placed in a diagonal position, the colored rings are replaced by colored curves which are reminiscent of hyperbolas (Fig. 98).

Colored interference figures can be observed only with sufficiently thick (0.5 mm or more) plates, sufficiently convergent light, and large magnification. When the plates are thin and the magnification not high enough, only a small part of the interference pattern is observed. Under these conditions, in sections perpendicular to the optic axis, one sees a diffuse dark cross on a grey background, while in sections at a large angle to the optic axis, the plate will simply be extinguished or translucent when it is rotated about its normal, just as in the case of parallel light. It is important to remember that when one uses a thick plate and stronger objectives, a large number of colored rings can be seen so that they appear to be much narrower. With very thick

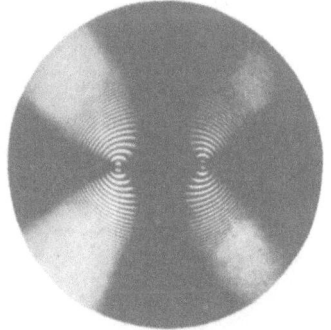

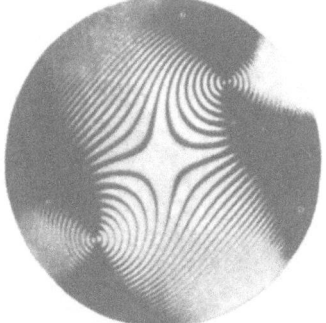

Fig. 94. Interference pattern obtained with a biaxial crystal placed between crossed Nicol prisms and observed in convergent light. The acute bisectrix is perpendicular to the plate. The vibration directions coincide with the vibration directions in the Nicol prisms—direct position. A dark cross and colored Cassini ovals are obtained in white light. A dark cross with dark and bright equally colored Cassini ovals is observed in monochromatic light. The exits of the optic axes can be seen.

Fig. 95. Interference pattern obtained with a biaxial crystal plate after a rotation of 22° from the direct position about one of its normals. The crystal has a larger optic axial angle than the crystal in the previous drawing. The remaining conditions are the same (cf. caption under Fig. 94).

plates and even stronger objectives, the rings may turn out to be so close together that the eye cannot separate them. In such cases, only a black cross can be seen on a white background, but the cross is very black. In equally thick plates, the number of rings in the field of view will be larger in the crystal whose birefringent power is larger.

It is clear from the above discussion that the observed colored curves (isochromes) and dark arms (isogyres) are in good general agreement with the theoretical predictions. Small deviations of the observed curves from the theoretical are due to the optical apparatus used to observe the interference patterns.

Interference Patterns Obtained with Biaxial Crystals in Convergent Light. When the crystal plate is cut perpendicular to the acute bisectrix, and is placed between crossed Nicol prisms so that the optic axial plane coincides with the plane of vibrations in one of the

Nicol prisms, the interference pattern obtained with biaxial crystals in monochromatic light consists of dark and bright isochromes which are almost identical with the Cassini ovals, and a dark cross whose arms are parallel to the vibration planes in the Nicol prisms (Fig. 94). When the plate is rotated about a normal to it, the arms of the cross

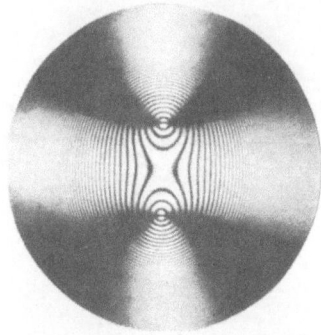

Fig. 96. Interference pattern obtained with a biaxial crystal in a diagonal position between crossed Nicol prisms.

Fig. 97. Interference pattern for a biaxial crystal placed between crossed Nicol prisms in convergent light. One of the optic axes is normal to the plate.

separate and form a pair of nonintersecting isogyres (Fig. 95), which are symmetrical with respect to the optic axial plane when the plate is in the diagonal position (Fig. 96). Points corresponding to the exits of the optic axes, where the final path difference is equal to $\frac{1}{2}\lambda$, are dark, just as in the case of uniaxial crystals. The first bright rings around these points correspond to a final path difference of λ, the first dark rings correspond to a path difference of $\frac{3}{2}\lambda$, the second bright rings to 2λ, and so on.

In biaxial crystals cut perpendicular to one of the optic axes, the interference pattern consists of rings which are not strictly circular, nor strictly concentric about the point of exit of the optic axis (Fig. 97). The rings are intersected by a single dark arm (isogyre). When the plate is rotated about one of its normals, the isogyre rotates in the opposite direction, which can be explained by the appropriate isogyre construction using an elliptical net on a sphere.

In the case of plates cut in other ways, the interference pattern is intermediate between the above special forms. When the plate is cut parallel to the optic axial plane, the interference pattern (Fig. 98) is almost identical with the interference pattern obtained with uniaxial crystals cut parallel to the optic axis.

The special property of biaxial crystals is that the isogyres observed with them always bend when the plate is rotated about one of its normals.

Determination of the Optical Sign of Uniaxial Crystals in Convergent Light. The simplest way of determining the optical sign is to use a plate cut perpendicular to the optic axis. A quartz wedge can be used for this purpose. On introducing the thin edge of the wedge above the crystal plate placed in a diagonal position, one observes a motion of the interference rings which is different in the different quadrants of the axial pattern (Fig. 99). The rings move toward the center in the two quadrants in which the sections of the optical indicatrices of the plate and the wedge are matched (i.e., the major axis of the section of the indicatrix of the crystal coincides with the major axis of the wedge) while in the other two quadrants in which the two are unmatched (n' of the crystal coincides with the n'' of the wedge) the rings move in the opposite direction.

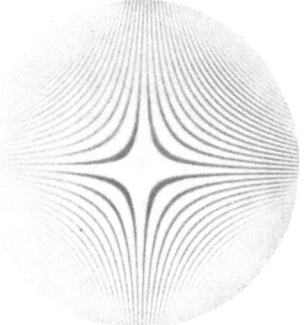

Fig. 98. Interference pattern obtained with a biaxial crystal placed between crossed Nicol prisms in convergent light. The plate is cut parallel to the optic axial plane.

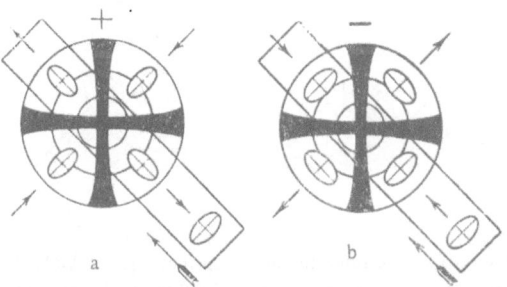

Fig. 99. Determination of the optical sign of a uniaxial crystal placed between crossed Nicol prisms in convergent light. The crystal plate is cut perpendicular to the optic axis. When the wedge is in a diagonal position and its thin end is introduced from the fourth quadrant of the field of view into the second, the interference rings move away from the center in the second and third quadrants, if the crystal is positive (a). If the crystal is negative, the opposite effect takes place (b).

105

The optical sign can also be determined with obliquely cut plates. In these cases, it is necessary to know where (right, left, above, etc.)

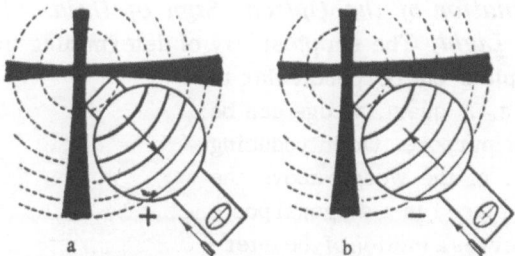

Fig. 100. Determination of the optical sign of a uniaxial crystal placed between crossed Nicol prisms in convergent light, using a quartz wedge. The crystal plate is cut at an angle to the optic axis. When the interference rings are as shown in the drawing, and the thin end of the quartz wedge is introduced into the field of view from the fourth quadrant into the second, the rings will move against the wedge if the crystal is positive (a). If the crystal is negative, the rings will move in the opposite direction (b).

the optic axis is located, remembering that the thin end of the dark arm is always directed toward the optic axis. Having established from

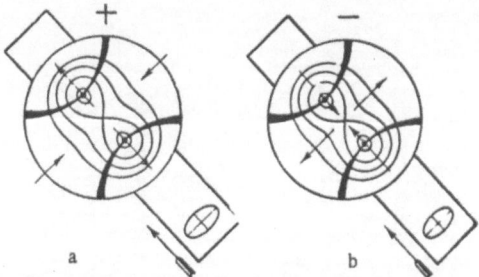

Fig. 101. Determination of the optical sign of biaxial crystals placed between crossed Nicol prisms in convergent light, using a quartz wedge. The plate is cut perpendicular to the acute bisectrix. The drawing shows the direction of motion of the rings when the thin end of the wedge is introduced from the fourth quadrant into the second. When the crystal is positive, the bands move away from the center (the exit of the acute bisectrix) in the second and fourth quadrants, and toward the center in the remaining quadrants (a). In the case of a negative crystal, the bands move in the opposite directions (b).

106

the position of the optic axis which of the four quadrants of the cross lies in the field of view, the optical sign can easily be determined by the method just described (Fig. 100).

Determination of the Optical Sign of Biaxial Crystals. The above method of determination of the optical sign of uniaxial crystals

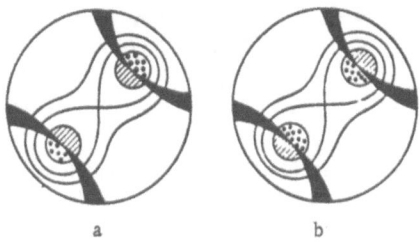

a   b

Fig. 102. The typical cases of dispersion of axes in orthorhombic crystals placed between crossed Nicol prisms in convergent light. The case $v_v >$ $> v_r$ is shown on the left (a). The case $v_v < v_r$ is shown on the right (b). Red color is indicated by dots, violet by dashes. It must be particularly noted that for $v_v > v_r$ the distance between the violet spots is smaller than between the red. For parallel Nicol prisms the opposite arrangement of colors is observed (cf. Fig. 51).

can also be used with some modifications in the case of biaxial crystal plates cut perpendicular to the acute bisectrix. In this case, when the quartz wedge is introduced, the colored curves will also move, but their motion will be complicated by the presence of two dark hyperbolas (in the diagonal position of the plate) instead of the cross, and two optic axes instead of one (Fig. 101). In positive crystals, the curves will move in the direction of the convex sides of the isogyres, and in negative crystals they will move in the opposite direction.

In oblique sections, one often observes only a single isogyre. In such cases, the optical sign can also be determined with the aid of a quartz wedge from the motion of the colored rings. It must only be recalled that the acute bisectrix lies in the convex side of the hyperbola, which is easily allowed for in view of the above discussion.

The dispersion of axes and bisectrices in convergent light is most strongly expressed in the position of the first order colors, which group themselves near the exits of the optic axes. In orthorhombic crystals, the colors are located symmetrically with respect to the two symmetry planes (Fig. 102). In monoclinic crystals, the colors

107

are located symmetrically with respect to the symmetry plane (two cases: Figs. 103 a, b) or symmetrically with respect to the twofold axis (Fig. 103 c). In triclinic crystals the colors are located asymmetrically (Fig. 104).

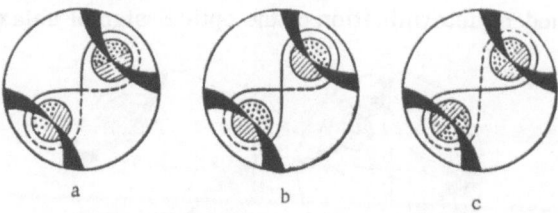

Fig. 103. Dispersion of bisectrices in monoclinic crystals placed between crossed Nicol prisms in convergent light. a) Horizontal; b) oblique; c) crossed.

Observation of Interference Patterns in a Crystal Sphere.

The best method for preliminary studies and demonstrations of the

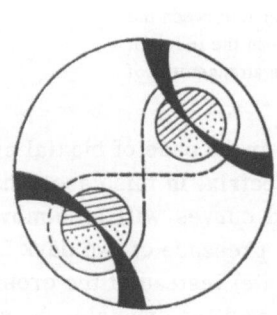

Fig. 104. Dispersion of bisectrices in triclinic crystals placed between crossed Nicol prisms in convergent light. General case (cf. Fig. 54).

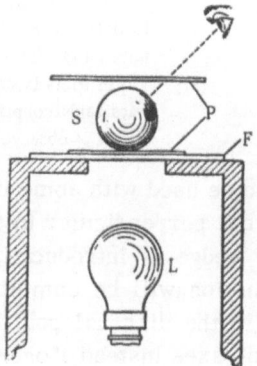

Fig. 105. Observation of the interference pattern in a crystal sphere. The sphere S is placed between two polaroids P. The frosted glass plate F, illuminated from below by the lamp L, serves as the source of light.

optical properties of a crystal, is to use a monocrystalline sphere. The observations are carried out in parallel light, while the interference pattern obtained looks like a pattern obtained in convergent light. This is due to the fact that the sphere acts both as the object under investigation and a collecting lens. The simple instrument shown in Fig. 105 may be used to study the optical phenomena in the sphere. It is convenient because the interference pattern obtained with the sphere can be observed through the upper polaroid directly by the eye at various angles. This means that the isochromatic surfaces can be looked at from all sides in space.

Another method of demonstrating interference patterns, using the above instrument, consists of the following. A crystal plate (mica, quartz, Rochelle salt, ammonium phosphate, gypsum, etc.) of sufficient size, is placed directly on the lower polaroid of the instrument. A glass sphere is then placed on the plate. The larger the sphere, the larger the interference pattern, and the more effective is the demonstration. The pattern is observed through the upper polaroid, which in this case is best held in the hand, so that the pattern can be conveniently looked at from all directions.

a   b   c

Fig. 106. Interference patterns observed in spherulite plates placed between crossed Nicol prisms in parallel light. a) Usual case; b) spherulite formed from uniaxial twisted filaments; c) spherulite formed from biaxial twisted filaments.

Interference Patterns in Spherulites. When the spherulites are in the form of thin, plane parallel sections, or crystallize out between two glass slides, and are placed between crossed Nicols in parallel light, the interference patterns obtained resemble the axial patterns which are obtained in convergent light with uniaxial crystal plates cut perpendicular to the optic axis. Three different kinds of interference patterns produced by spherulites can be distinguished.

1. The most frequently occurring form consists of a dark cross and a luminous field (Fig. 106 a). The difference between a spherulite cross and a cross obtained with a uniaxial crystal lies in the fact that the arms of the former come more clearly to a point at the center, and its edges are not curved. The cross is due to the direct extinction of the needle shaped crystals which form the spherulite and are arranged radially from its center. The direct extinction itself, i.e., extinction in the case when the crystal is placed so that its length is parallel to the vibration direction in one of the Nicol prisms, is due

109

to the fact that one of the symmetry axes of the crystal indicatrix is parallel to its length. It is easy to see that it is possible to obtain a cross on a luminous uniformly colored or uncolored field with spherulites formed from both axial and uniaxial crystalline needles.

2. The second type of interference pattern obtained with spherulites has the form of a dark cross with concentric dark and bright rings at different distances from each other (Fig. 106 b). The interference hue periodically increases and decreases along the radius from the center of the spherulite (while in the case of the axial figure obtained with a uniaxial crystal it continuously increases). This type of pattern is observed in spherulites consisting of uniaxial twisted crystalline filaments with optic axes perpendicular to them. Clearly, at those points on such a filament where the optic axis is perpendicular to the plane of the specimen, the interference pattern should be dark, while when the optic axis is in the plane of the specimen it should show a maximally high hue. The distance between nearest dark points should be equal to each other and also equal to half the pitch of a double-thread screw.

3. The third known form of interference patterns obtained with spherulites consists of a dark cross and concentric dark rings which are at a distance alternately large and small from each other (Fig. 106 c). This form occurs when the spherulite consists of equally twisted biaxial monocrystalline filaments with the optic axial plane perpendicular to each of them. Under these conditions, the optic axes in a uniformly twisted filament should be alternately perpendicular to the plane of the specimen and thus produce a sequence of extinguished regions with interference colors between them. If the angle between the optic axes is 90°, the distance between the rings is constant. Since, however, the angle between the optic axes is usually unequal to 90°, the distance between the dark rings are alternately large and small.

ROTATION OF THE PLANE OF POLARIZATION

The Main Phenomenon. The phenomenon of double refraction, which was described in the previous chapter, is observed in a pure form, uncomplicated by other phenomena, only in transparent uncolored crystals which do not rotate the plane of polarization.

Iceland spar is such a crystal. We know that a calcite plate cut along planes perpendicular to the optic axis remains extinguished when it is rotated about the optic axis and observed in a parallel beam of white or monochromatic light between crossed Nicol prisms. It is important to note that under these conditions both thin and thick plates remain extinguished.

Under similar conditions, a quartz plate cut perpendicular to the optic axis will remain extinguished only if it is sufficiently thin. As the thickness increases, and if white light is used, the plate will transmit light and will become colored. If the plate is then rotated about the optic axis, the intensity of the transmitted light and its hue remain unchanged. Experiment shows that to reestablish extinction in monochromatic light, it is necessary to rotate the analyzer through an angle about the optic axis, while in white light, the extinction cannot be reestablished by any rotation of the analyzer about the axis of the polarizing instrument. The above phenomenon was discovered in quartz by the French scientist Arago in 1811 and is known as rotation of the plane of polarization.

Sign of Rotation. If instead of a quartz plate one takes a quartz wedge, one plane of which is cut perpendicular to the optic axis (which should not be confused with the quartz wedge cut parallel to the optic axis and described earlier), and if the wedge is introduced into the polarizing instrument so that this plane is normal to the beam of light, then it is found that the angle of rotation of the analyzer, necessary to extinguish the monochromatic light transmitted by the wedge, is

111

proportional to the appropriate thickness of the wedge. At the same time, the sign of the rotation may be determined, i.e., whether the analyzer should be rotated to the right or to the left in order to reestablish extinction with gradually increasing wedge thickness. Left-handed rotation (counterclockwise) is normally considered to be positive and right-handed rotation negative.

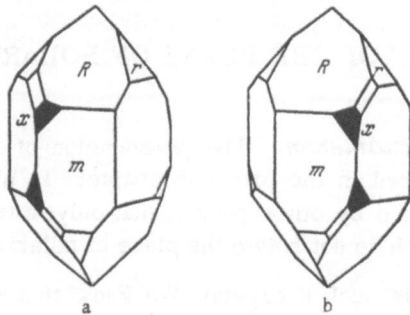

Fig. 107. Ideal forms of left-handed (a) and right-handed (b) quartz. If the face *m* of the prism is placed so that the face *R* of the main rhombohedron is above it, then in left-handed quartz the faces *x* of the trapezohedron will be to the left of *m* and are blackened in the diagram. In right-handed quartz the faces *x* will be on the right of *m* under similar conditions.

If during the observation of the rotation of the plane of polarization, the quartz plate is turned over, then provided the position of the light source and the eye remains unaltered, no change in the observed phenomenon will take place. It follows that the sign of the rotation does not depend on the direction of propagation of the ray along the optic axis within the crystal.

Connection Between the Sign of Rotation and the Enantiomorphism of Crystals. Herschel (1821) was the first to point out the close connection which exists between the sign of the rotation and the morphological enantiomorphism of quartz crystals. It turns out that quartz crystals which according to morphological criteria are right-handed have positive rotation, and vise versa. We recall that quartz crystals are called right-handed if the faces *x* of the trapezohedron are on the right, below the faces *R* of the main rhombohedron, and left-handed if the faces *x* lie on the left, below the *R* faces (Fig.

107). We shall see later (p. 139), that right-handed (left-handed) quartz crystals have right-handed (left-handed) rotation only in directions close to the optic axis. The sign of the rotation is opposite in directions perpendicular to the optic axis, or directions close to them.

The Symmetry of Quartz in Relation to the Phenomenon of Rotation of the Plate of Polarization. Since the rotation of a quartz plate through any angle about the optic axis, and its rotation through 180° about any normal to the optic axis, do not lead to any changes in the observed phenomenon of rotation of the plane of polarization, and since we can distinguish between right-handed and left-handed rotation, it follows that in relation to this phenomenon we may ascribe

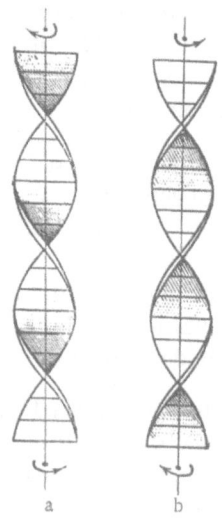

Fig. 109. Model of the rotation of the plane of polarization. a) A ribbon twisted to the right (left-handed screw) representing right-handed rotation of the plane of polarization; b) a ribbon twisted to the left (right-handed screw) representing left-handed rotation of the plane of polarization.

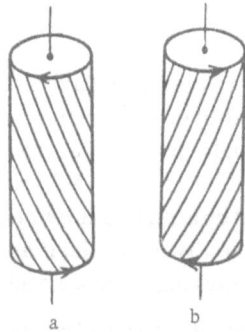

Fig. 108. Examples of figures having an infinite-fold symmetry axis and an infinite number of transverse twofold axes. a)Left-handed;b)right-handed.

a ∞ : 2 macrosymmetry to quartz, i.e., the symmetry of a twisted cylinder (Fig. 108). The latter is characterized by an infinite-fold axis, an infinite number of transverse twofold axes, and the absence of planes and center of symmetry. It is known that the morphological symmetry of quartz is described by the 3 : 2 group (threefold axis, and three twofold axes perpendicular to it). This group is a subgroup of the ∞ : 2 group. We thus see that in this case also, the important crystallophysical law frequently mentioned above (pp. 55, 56, 60, etc.),

113

which states that the morphological symmetry of a crystal either coincides with the symmetry of the property under investigation or is subordinate to it, is confirmed again.

The Twisting of the Plane of Polarization. It is clear from the above discussion that inside a crystal, the phenomenon which we have called the rotation of the plane of polarization is in fact not a rotation but a twisting of this plane (Fig. 109). If one cuts a ribbon out of this plane in the direction parallel to the ray, then when the ribbon is twisted, a double-thread screw is formed. A left-handed thread will then correspond to a right-handed rotation of the plane of polarization, and vice versa.

The symmetry of such a screw is described by the linear symmetry group $2 \cdot \infty : 2$. It includes an infinite-fold screw axis of symmetry, a simple twofold axis which coincides with the ∞-axis, an infinite number of transverse twofold axes, and a translation axis which coincides with the ∞-axis and which has an elementary transport equal to half the screw pitch. The macroscopic symmetry of this screw is described by the group $2 : 2$.

Earlier we agreed to imagine linearly polarized rays as infinite sinusoids, or sections of it, which move uniformly in a fixed plane. In the phenomenon of rotation of the plane of polarization we are dealing, inside the crystal, with a beam of light polarized in a special way (along a screw surface) which may be represented by a sinusoid twisted about its axis and moving uniformly over the fixed twisted plane and in the direction of its ∞-axis, similarly to a nut moving over a fixed screw.

Specific Rotation and Rotary Dispersion. It was already pointed out that the angle of rotation of the plane of polarization of monochromatic light is proportional to the thickness of the plate so that

$$\rho_d = \rho \cdot d. \tag{92}$$

The coefficient of proportionality ρ between the angle of rotation of the plane of polarization ρ_d and the thickness of the plate d is known as specific rotation. In crystals, specific rotation is usually expressed in degrees per millimeter of thickness. Specific rotation is different for different crystals and is a function of temperature and wavelength. The dependence of ρ on λ is known as rotary dispersion. In the case of normal dispersion, the specific rotation increases as the wavelength decreases. There is complete analogy between rotary dispersion and refractive dispersion. Anomalous rotary dispersion is also known to occur.

114

In 1814 Biot found that specific rotation is inversely proportional to the square of the corresponding wavelength so that

$$\rho = \frac{A}{\lambda^2} . \tag{93}$$

More exactly, the dependence of ρ on λ may be expressed by the series

$$\rho = \frac{A}{\lambda^2} + \frac{B}{\lambda^4} + \cdots \tag{94}$$

As an example, we give the values of specific rotation of quartz along the optic axis for seven Fraunhofer lines (Table V).

TABLE V

Specific Rotation of Quartz

Line	λ,mμ	ρ°/mm	Line	λ,mμ	ρ°/mm
B	686.7	15.55	F	486.1	32.69
C	656.3	17.22	G	430.8	42.37
D	589.3	21.67	H	396.9	50.98
E	527.0	27.46			

Specific rotation of right-handed and left-handed quartz is numerically the same.

In the case of anomalous dispersion, specific rotation depends not only on the frequency of the transmitted light but also on the natural vibration frequency of the molecules of the substance under consideration.

Interference Hue Due to the Rotation of the Plane of Polarization. If a quartz plate, or any other crystal plate which produces rotation of the plane of polarization, is placed between crossed Nicol prisms and is observed in parallel white light, it is found to assume an interference hue. Experiment shows that this hue depends not only on the thickness of the plate but also on the rotation of one of the Nicol prisms about the axis of the beam of light. However, if the plate itself is rotated about its optic axis then neither the intensity of the transmitted light nor the hue are found to change. This is explained as follows.

Suppose that in the light transmitted by the first Nicol prism I, which transmits vibrations along the vertical (Fig. 110), red (r), orange (o), etc., rays leave the prism with vibrations at different

115

angles to the vertical. It is clear that the second Nicol prism II, which is crossed with the first, will freely transmit only those rays whose vibrations are parallel to the horizontal, and will completely extinguish all rays whose vibrations are parallel to the vertical. The remaining rays will be transmitted to a greater or lesser extent depending how near they are to the horizontal or the vertical. In the concrete case shown in the drawing, the yellow-green rays will be completely transmitted, while violet rays will be completely extinguished. Accordingly, the transmitted light will have a mixed yellow-green hue.

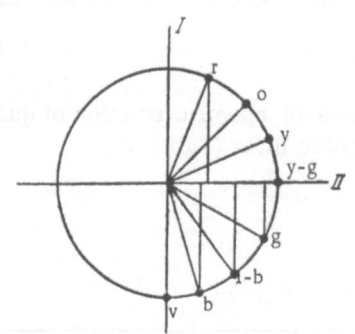

Fig. 110. Diagram illustrating the phenomenon of rotary dispersion. Vibration directions corresponding to different colors are rotated through different angles.

When one of the Nicol prisms is rotated in clockwise or counterclockwise direction, the mixed hue should change in accordance with the composition of the transmitted and absorbed rays. In the case shown in the drawing, when the analyzer is rotated in the clockwise direction green, light blue, dark blue, violet, etc., rays will be transmitted in succession.

We therefore obtain the following rule for the determination of the sign of the rotation. If, when the analyzer is rotated in a clockwise direction, the interference hue increases, then the crystal under consideration rotates the plane of polarization in the clockwise direction. If the hue is lowered under similar conditions, then the crystal rotates the plane of polarization in the counterclockwise direction. Since the rotation of the analyzer in the clockwise direction is equivalent to the rotation of the polarizer through a similar angle but in the counterclockwise direction, it follows that the converse rule must be used in the case of the polarizer.

Let us now consider the change in the hue of the plate as the thickness increases. Since as the thickness of the plate increases, the angle of rotation of the plane of polarization also increases for all the rays, it follows that the analyzer will no longer transmit completely only the yellow-green rays but, instead, it will transmit yellow rays, and if the thickness of the plate is increased further, orange rays will be transmitted, then red, and so on. As a result of the increase in the thickness of the plate, the interference hues will vary

116

(increase) in the same order along the color scale as in the case of a doubly refracting plate of increasing thickness. Experiment shows that in order to obtain a given interference hue due to the rotation of the plane of polarization, the required plate thickness is much greater than that necessary to obtain the same hue due to double refraction. As an example, let us mention the fact that the necessary thickness of a quartz plate, cut perpendicular to the optic axis and placed between crossed Nicol prisms, is 7.5 mm in the case of violet rays, while the same hue is obtained with a quartz plate cut parallel to the optic axis with a thickness a hundred times smaller (0.064 mm).

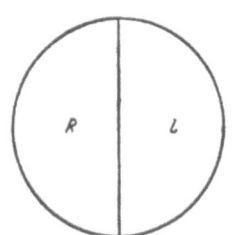

Fig. 111. Biquartz. A plate of right-handed quartz is cemented to a plate of left-handed quartz L. The two plates have the same thickness (7.5 mm) and are cut perpendicular to the optic axis.

We know already that the transition from crossed Nicol prisms to parallel prisms always leads to the complementary hue. This is also observed in the case of rotation of the plane of polarization, since the interference hue in this case is due to a path difference, which is different for rays of different wavelength, and the reduction of the vibration directions of all the rays to a single plane.

The Biquartz Plate. The violet tint of a quartz plate 7.5 mm thick, placed between crossed Nicol prisms (or 3.75 mm thick and placed between parallel Nicol prisms) is called violet sensitive. This name is due to the fact that a small rotation of one of the Nicol prisms about the axis of the polarizing instrument changes this tint into red or blue. The sensitive tint of a quartz plate can therefore be used to set the Nicol prisms accurately in their crossed position, and to determine small values of rotation of the plane of polarization in plates placed in series with the quartz plate. An even more convenient plate for these purposes is the biquartz plate which consists of two plates of levo- and dextro-quartz stuck together as shown in Fig. 111 and having the appropriate sensitive tints. A very small rotation of one of the Nicol prisms, or the superposition on the biquartz of a plate which rotates the plane of polarization, produces an opposite change in the tint of the two halves of the biquartz which can be easily detected.

Circular Polarization. So far, by a polarized ray we always understood a linearly polarized ray. We shall see later (p. 145) that this is not the only kind of polarization. Here, we shall briefly consider circular polarization, since this will be necessary to the explanation of the phenomenon of rotation of the plane of polarization. In order to

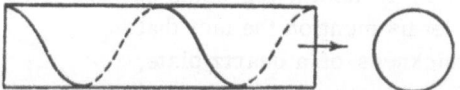

Fig. 112. Model of circularly polarized ray. The spiral drawn on the circular cylinder moves uniformly in the direction of the axis of the cylinder.

obtain a clear picture of what is meant by circular polarization, it is sufficient to consider the following similarities and differences between linearly and circularly polarized rays.

Monochromatic linearly polarized ray	Monochromatic circularly polarized ray
Light vectors are transverse to the ray and execute harmonic vibrations.	Light vectors are transverse to the ray and rotate uniformly about the ray.
The end points of the light vector form a sinusoid at any given time.	The end points of the light vectors form a screw thread at any given time (Fig. 112).
A linearly polarized beam may be represented by a model consisting of a sinusoid moving uniformly in the direction of its axis.	A circularly polarized ray may be represented by a model consisting of a screw thread moving uniformly in the direction of its axis, or equivalently, uniformly rotating about its axis.
A linearly polarized ray is completely defined if the amplitude of the light vector A, the wavelength λ and the velocity v are given.	A circularly polarized ray is completely defined if the amplitude of the light vector A, the wavelength λ and the velocity v are given.
Linearly polarized rays do not possess a rotation sign.	Circularly polarized rays can differ in the sign of the rotation.

118

Rotation of the Plane of Polarization as Double Refraction of Circularly Polarized Rays (Fresnel's Theory). Consider two equal vectors A' and A'' drawn from a common origin, rotating in a given plane with equal velocities and in opposite directions. It is clear

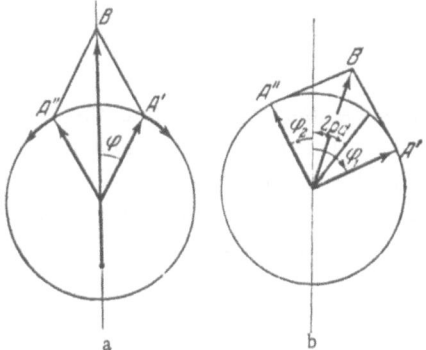

Fig. 113. Diagram illustrating the phenomenon of rotation of the plane of polarization. Decomposition of the harmonic vibrations represented by the vector B into two uniformly revolving vectors A' and A'' which rotate in opposite directions but with the same frequency (a). Addition of the vectors A' and A'' rotated in opposite directions and through different angles.

from Fig. 113 a that the resultant vector B will then execute linear harmonic vibrations with an amplitude $2A'$, and a phase φ equal to the phase of each of the component vectors. It may therefore be concluded that any linearly polarized ray may be looked upon as consisting of two circularly polarized rays of opposite sign but traveling in the same direction.

In 1825 Fresnel gave the following explanation of the rotation of the plane of polarization. When a linearly polarized monochromatic ray enters a quartz plate, it is resolved into two circularly polarized rays of opposite sign, which travel inside the crystal along the same direction but with different velocities. As a result, when the two rays leave the crystal there is a phase difference $\Delta = \varphi_1 - \varphi_2$. between them. The presence of this phase difference means that the linearly polarized ray which is formed as a result of the recombination of the two circularly polarized rays is polarized in a plane which is at an angle ρ_d to the initial plane of polarization.

It is clear from Fig. 113 b that this angle is equal to one half of the phase difference, i.e.,

119

$$\rho_d = \frac{\Delta}{2} \cdot \tag{95}$$

Using equation (68), which holds for waves of any form, we have

$$\rho_d = \frac{\pi d \, (n' - n'')}{\lambda}$$

and finally

$$\rho = \frac{\pi \, (n' - n'')}{\lambda} \cdot \tag{96}$$

This formula may be used to calculate the circular birefringent power $n' - n''$ of quartz from the easily measurable quantities ρ and λ (ρ and λ in equation (96) should of course be expressed in consistent units, for example, in radians per centimeter and in centimeters, respectively). Table VI gives the result of such a calculation for seven Fraunhofer lines.

TABLE VI

Circular Birefringent Power of Quartz ($n' - n''$) Along the
Optic Axis for Seven Fraunhofer Lines

Line	λ, mμ	$n' - n''$	Line	λ, mμ	$n' - n''$
B	686.7	0.000059	F	486.1	0.000088
C	656.3	0.000063	G	430.8	0.000101
D	589.3	0.000071	H	396.9	0.000111
E	527.0	0.000080			

Experimental Confirmation of the Fresnel Theory. The explanation of the rotation of the plane of polarization which was given by Fresnel can be verified by an experiment which was also carried out by Fresnel. In this experiment, Fresnel prepared a compound quartz prism consisting of three separate prisms as shown in Fig. 114. The

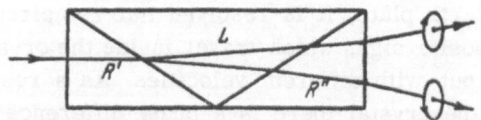

Fig. 114. Decomposition of a linearly polarized or natural ray into two circularly polarized rays by means of a Fresnel prism. Fresnel's prism consists of three prisms R', R'', and L. Prisms R', R'' are made from right-handed quartz and prism L from left-handed quartz. The rotation of the electric vector in the two rays leaving the prism is in opposite directions.

end prisms R' and R'' were made from quartz of the same sign, and the third prism L was made from quartz of the opposite sign. The optic axes of the three prisms are all parallel to the length of the prism. A linearly polarized light is incident normally on the end face of the prism R' and enters it without refraction. Inside the prism it is decomposed into two circularly polarized rays traveling in the same direction. Since these two rays have different refractive indices they are unequally refracted at the boundary between R' and L. If the prism R' is made of right-handed quartz, then the ray with right-handed rotation will have a larger velocity, and correspondingly, a lower refractive index n'', compared with the other ray which has a lower velocity and a larger refractive index. On entering the prism L the two rays interchange their roles: the one which in R' had the lower velocity will travel through L with a larger velocity, and vice versa. As a result, when the ray with the refractive index n' enters the medium which is less refractive for it, it should be bent away from the normal to the boundary between the two prisms, while the other ray, for which the prism L is more refracting, should be bent toward the normal. In other words, the two rays should diverge at the boundary between R' and L. This divergence will be increased, for the same reason, at the boundary between L and R'', and again when the rays leave the final prism. The two rays should then be circularly polarized in opposite directions. This is entirely confirmed by the experiment.

We began with the assumption that a linearly polarized light is incident on the Fresnel prism. Our discussion will still hold in the case of unpolarized light, since unpolarized light consists of polarized quanta (p. 21).

Comparison Between the Twisted Plane Model and the Double Refraction of Circularly Polarized Rays. When we say that white light consists of monochromatic rays of different frequency we mean that it can be decomposed into rays of different color. In this sense one could say that any vector consists of the components into which it can be resolved. However, the statement that a wooden plank used to make a shelf consists of shavings, sawdust and the shelf, cannot be seriously defended. We thus see that the problem of what does white light really consist of, and whether it consists of linearly or circularly polarized light, is not as simple as might seem at first sight. Bearing this in mind, we sometimes have to reconcile ourselves to this duality (and sometimes even multiplicity) in the explanation of the various special cases.

If one accepts Fresnel's suggestion that linearly polarized mono-chromatic light consists of two circularly polarized rays moving in the same direction, then it is also necessary to accept that these rays, which have different velocities within the crystal, and consequently different wavelengths, should form group waves when they interact with each other (Fig. 15). The velocity of the group waves is given by the formula

$$V = \frac{\lambda' v'' - \lambda'' v'}{\lambda' - \lambda''}$$

which was given earlier (equation (18)) and is found to be zero since the frequency of the vibrations (the number of rotations of the light vector per second) is the same in the two rays, i.e.,

$$v = \lambda' v'' = \lambda'' v'.$$

The group wavelength, on the other hand, is given by equation (16):

$$L = \frac{\lambda' \lambda''}{\lambda' - \lambda''}.$$

The twisted plane model described above (Fig. 109) also represents the stationary group waves which are formed as a result of the inter-action (geometrical addition of the electric vectors) of the two circularly polarized waves, which according to Fresnel, should be prop-agated along the same direction but with different velocities inside the crystal. It must be emphasized that these group waves are sta-tionary only in the sense that their normal velocity V is zero. This does not mean that their tangential velocity is also equal to zero, or that other motion is absent. The important fact is that the only pos-sible motion in the group waves can take place only on the above sta-tionary screw surface.

It was already pointed out that the wavelength L of the group waves, which is equal to the screw pitch in the twisted plane, can be calcu-lated from the general formula (16) which in our case assumes the form

$$L = \frac{\lambda' \lambda''}{\lambda' - \lambda''} = \frac{\lambda_0}{|n'' - n'|}$$

or, if one uses the specific rotation ρ,

$$L = \frac{180°}{\rho} \text{mm}.$$

For the solar D line in quartz, $L = 8.3$ mm, i.e., the group wave-length is in this case 1400 times larger than the wavelength ($\lambda_D =$ $= 589.3$) of the corresponding rays in a vacuum.

It follows from the above discussion that, from the mathematical point of view, rotation of the plane of polarization may be looked upon either as the twisting of the plane of polarization or as the decomposition of linearly polarized rays into two circularly polarized rays propagated in the same direction but with different velocities. Moreover, we are concerned strictly with the rotation of the plane of polarization and not the decomposition of a single ray into two rays moving in different directions, i.e., we are concerned with what goes on in the R' part of the Fresnel prism and not with the L and R'' parts where rotation of the plane of polarization does not take place.

Mirror Isomerism of Molecules and Its Connection with Rotation of the Plane of Polarization. Just as the crystals of a given substance, for example quartz crystals, can in many cases exist in two enantiomorphic modifications, so also the molecules of a number of substances, for example the molecules of tartaric acid, can exist in two forms, namely, right-handed and left-handed. Substances consisting of only left-handed or only right-handed molecules, and also substances in which the concentration of molecules of a given enantiomorphic modification predominates over the concentration of other molecules, can rotate the plane of polarization not only in the crystalline state but also in the amorphous state (liquid or vitreous). The phenomenon of enantiomorphism in the case of molecules is known as optical or mirror isomorphism, and the corresponding substances have (unfortunately) become known as optically active.

Pasteur (1860) was the first to suggest that the reason for the optical activity of liquids is molecular dissymmetry. It is important to note that various authors have used the concept of dissymmetry in different senses. By a dissymmetry figure Pasteur understands a figure which cannot be made to coincide with its mirror image by a simple superposition. The human hand is an example of a disymmetric figure: when the right-hand is reflected in the mirror its figure is transformed into the figure of a left-hand but the two cannot be made to coincide by a simple superposition. Dissymmetry in Pasteur's sense is characterized by the absence of symmetry elements of the second kind (symmetry planes, mirror axes, symmetry center and planes of sliding reflection). It follows that dissymmetry should not be confused with its special case, namely, assymmetry, i.e., the total absence of symmetry. This means that finite dissymmetric figures include not only class *1* figures but also all figures of the purely axial classes, both crystallographic and noncrystallographic (containing fivefold, and higher than sixfold axes), i.e.,

$$1, 2, 3, 4. \quad . \quad . \quad \infty$$
$$2{:}2, 3{:}2, 4{:}2 \quad . \quad . \quad \infty : 2$$
$$3/2, 3/4, 3/5, \infty/\infty.$$

Van't Hoff (1875) developed Pasteur's idea and suggested that the optical isometry of molecules may be reduced to two possible waves of locating in them four different atoms or groups of atoms at the vertices of a tetrahedron having a carbon atom at its center (Fig. 115).

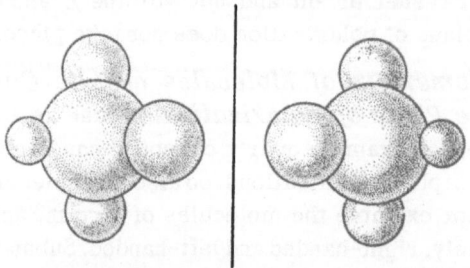

Fig. 115. Right-handed and left-handed arrangement of four different groups of atoms around a central carbon atom. Mirror isomerism of molecules.

Each such tetrahedron is an assymmetric figure since one such tetrahedron cannot be made to coincide with another by simple superposition. It is clear from Fig. 115 however that they can be made to coincide for example by reflection in a plane or, for a different mutual disposition, by inversion at a point.

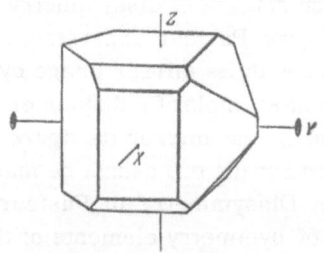

Fig. 116. Ideal form of a tartaric acid crystal. A twofold axis coincides with the Y axis.

We thus see that Van't Hoff narrowed down Pasteur's idea by reducing the real cause of optical activity of amorphous bodies, namely, molecular dissymmetry to its special case, i.e., molecular assymmetry, but this did not prevent him from making several brilliant discoveries in connection with optical activity of organic substances and thus lay down the foundations of stereochemistry.

Liquids which contain equal quantities of right-handed and left-handed molecules (racemic mixtures $R + L$) do not rotate the plane of polarization. Pasteur used salts of tartaric acid (Fig. 116) to show that when such liquids crystallize, equal quantities of right-handed and left-handed crystals are produced (according to morphological criteria). He further discovered that when dextrorotary (levorotary) crystals are dissolved in water, a dextrorotary (levorotary) solution is obtained. It was later established that dissymmetric crystals obtained by Pasteur's method from racemic mixtures are optically active, the optical activity being of one sign for right-handed crystals and of another for left-handed crystals.

In a number of cases, R and L molecules form a new compound RL. Such anticompounds, for example antitartaric acid, are inactive both in the liquid and the crystalline states, although from symmetry considerations optical activity in the crystalline state is not impossible.

Substances which are active in the liquid state but inactive in the crystalline state do not, and apparently cannot, exist. However, it is definitely established that active crystals can lose their activity when dissolved or heated. This is the case, for example, in quartz, which in its vitreous state (fused quartz) does not rotate the plane of polarization. The optical activity of such crystals is determined not by the dissymmetry of molecules which does not exist in such crystals, but the dissymmetry in the crystal structure, i.e., dissymmetric disposition of atoms (ions) in space.

Specific Rotation of the Plane of Polarization as a Directed Quantity. The above discussion already suggests that specific rotation of the plane of polarization ρ is a directed quantity, i.e., it is measured along a given direction and is, in general, dependent on direction. Just as any directed quantity, specific rotation can conveniently be represented by a segment of a straight line having a length proportional to $|\rho|$ and a direction coinciding with the direction in which this quantity is measured. However, specific rotation is neither a polar nor an axial vector. This is clear from the following.

It was pointed out above (p. 18, Fig. 13) that polar vectors can be represented by a straight arrow which represents all the important characteristics of a polar vector, namely, its numerical magnitude, its direction, and, very important, its $\infty \cdot m$ symmetry. An important property of any polar vector is that it will change its sign when its direction is reversed by any method. This means that a polar vector will change its sign on rotation through 180° about any of its normals, on reflection in a normal plane, and on inversion at a point (the center of the arrows). We know, however, that ρ does not change its sign when

125

the crystal plate, or the segment representing the quantity ρ, is rotated through 180° about one of its normals. It follows that ρ is not a polar vector. This is also indicated by the fact that straight arrows, which are figures having symmetry planes, cannot differ from each other in being right-handed or left-handed, while the quantity ρ can be either right-handed or left-handed.

In order to understand why ρ cannot be an axial vector either, we recall that a rational representation of any axial vector is a segment of a straight line with a single circular arrow (p. 18, Fig. 13), since this representation reproduces all the important properties of an axial vector, namely, its numerical magnitude, its direction, and its $\infty : m$ symmetry. In particular, this figure shows that an axial vector does not change its sign when it is reflected in a symmetry plane normal to it, or when it is inverted at its center of symmetry, but it will change its sign when it is rotated through 180° about any normal to it. The latter indicates that specific rotation is not an axial vector.

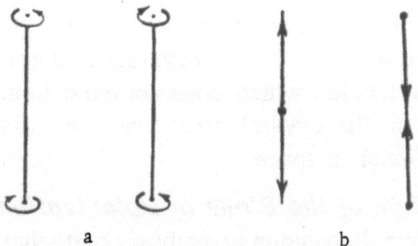

Fig. 117. Special cases of tensors. a) Left-handed and right-handed axial tensors; b) positive and negative polar tensors.

This is also indicated by the fact that segments of a straight line with one circular arrow are figures having a transverse symmetry plane, and hence cannot differ from each other by being right-handed or left-handed, while optical rotation can differ in this respect.

It follows from the above discussion that rotation of the plane of polarization is not, in fact, a vector. It is easy to verify that the only correct way of representing this quantity is by a segment of a straight line with two circular arrows at its ends, the arrows having different directions, but appearing as identical to an observer looking along the segment from either end (Fig. 117 a).

It is clear from the drawing that a segment of a straight line with two circular arrows has a $\infty : 2$ symmetry, and since it is a figure

126

which does not have symmetry elements of the second kind, it may be represented by two enantiomorphic forms, namely, right-handed and left-handed.

Quantities like specific rotation, which can be represented by a segment of a straight line with two oppositely directed circular arrows at the ends of the segment, are known as tensors, and in particular, axial tensors. The axial tensor ρ representing rotation of the plane of polarization differs from the corresponding polar tensors, such as the tensor (coefficient) of thermal expansion of crystals (Fig. 117 b), by their $\infty : 2$ symmetry. The corresponding polar tensor, e.g., the tensor of thermal expansion of crystals, has the $m \cdot \infty : m$ symmetry of an ellipsoid of revolution.

Dependence of Specific Rotation of the Plane of Polarization on Direction. We know from the foregoing discussion that the magnitude of specific rotation is, in general, different for different directions, but when a given direction is replaced by an opposite one, the specific rotation remains unaltered. This means that, firstly, ρ is a function of the direction cosines c_1, c_2, c_3 (p. 47), and secondly, this function is an even function, since replacement of a given direction by an opposite one leads to a change in the sign of all the three direction cosines, and the function itself can remain unaltered only if it is a function of even powers of the direction cosines. If we limit our attention to second powers only, the required dependence can be described following Chipart (1904) by the following equation

$$\rho = \rho_{11}c_1^2 + \rho_{22}c_2^2 + \rho_{33}c_3^2 + 2\rho_{12}c_1c_2 + 2\rho_{23}c_2c_3 + 2\rho_{31}c_3c_1, \tag{97}$$

where ρ_{ik} are constants. It is shown in tensor calculus that when the set of coordinates is suitably chosen, this equation may be reduced to the very simple form

$$\rho = \rho_1 c_1^2 + \rho_2 c_2^2 + \rho_3 c_3^2, \tag{98}$$

where ρ_1, ρ_2, ρ_3 are the principal constants. Their physical meaning is very simple. If a ray of light is directed along the X_1 axis then $c_1 = 1$, $c_2 = c_3 = 0$ and

$$\rho = \rho_1.$$

It follows that ρ_1 is the specific rotation along the first principal axis. The other constants have analogous meanings. Negative values of these constants correspond to right-handed rotation and positive values to left-handed rotation. Equation (98) shows that the specific rotation of a crystal in any given direction can be calculated (for rays of a given frequency) from the three principal values of the specific rotation of the crystal for these rays, in a way similar to that used, for example, in the calculation of the thermal expansion coefficient of a crystal in any given direction using the three principal expansion coefficients.

Gyration Surfaces. In the literature, this name is given to surfaces whose radius vectors are proportional to the square root of the absolute magnitude of the reciprocal of the specific rotation, i.e., $\sqrt{\frac{1}{|\rho|}}$. Such surfaces are not useful as models illustrating the dependence of specific rotation on the direction in a crystal, since they do not indicate the sign of the rotation nor the actual symmetry of the phenomenon, and since rotation of the plane of polarization itself is only represented by it in a somewhat conventional form. It must be added that some of these surfaces cannot be constructed at all, since they contain infinitely distant points. By gyration surfaces we shall understand those surfaces whose radius vectors are proportional to the corresponding values of the quantity ρ. Since, however, these radius vectors can be either positive or negative, we shall agree that those parts of the gyration surface which correspond to positive radius vectors will be drawn white and those which correspond to negative radius vectors will be drawn black. The construction of the gyration surfaces is not difficult, and requires the knowledge of the numerical values of the constants ρ_1, ρ_2, ρ_3 so that ρ can be calculated for all the possible directions when the direction cosines are given.

1. In cubic crystals all the three principal values of the specific rotation should be equal both in absolute magnitude and in sign:

$$.\rho_1 = \rho_2 = \rho_3.$$

This follows from the fact that any cubic crystal has among its symmetry axes three mutually perpendicular, and equal to each other, twofold or fourfold axes, i.e., three axes which transform into each other by ordinary symmetry operations. In the case of cubic crystals, equation (98) assumes the form

$$\rho = \rho_1 \left(c_1^2 + c_2^2 + c_3^2 \right) = \rho_1. \tag{99}$$

The corresponding surface is a sphere which is white for positive and black for negative ρ. Each of these spheres has the symmetry of the ∞/∞ group which contains an infinite number of infinite-fold symmetry axes oriented in all directions, and does not contain symmetry elements of the second kind (p. 123) at all. The absence of these elements is due to the fact that, in each of the two spheres, all the diameters are segments, of twisted straight lines (Fig. 117): in the white sphere (left-handed rotation) they are twisted along a right-handed screw and in the black sphere (right-handed rotation) they are twisted along a left-handed screw.

2. In uniaxial crystals two of the three constants ρ_1, ρ_2, ρ_3 should be equal to each other. If the third coordinate axis X_3 is taken as the principal axis of symmetry then ρ_1 and ρ_2 will be equal. However, in crystals this equality can be of two kinds: ρ_1 can either be directly equal to ρ_2, i.e., equal in absolute magnitude and in sign, or it can be

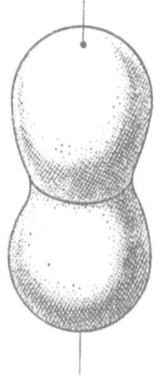

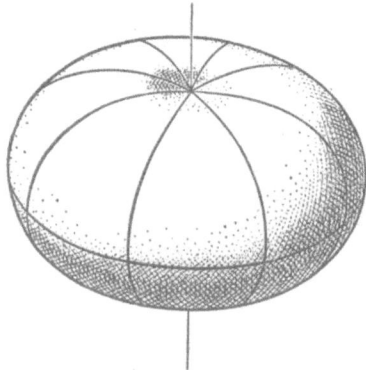

Fig. 118. Left-handed gy-
ration surface of uniaxial
crystals. Prolate surface
of revolution.

Fig. 119. Left-handed gyration surface of
uniaxial crystals. Oblate surface of revolution.

antiequal, i.e., equal in absolute magnitude but opposite in sign. In the first case, the specific rotation is identical along X_1 and X_2 axes, and in the second case, the specific rotation is right-handed along one of the axes and left-handed along the other. It follows that the relation

$$|\rho_3| \neq |\rho_2| = |\rho_1|$$

holds in uniaxial crystals.

Let us consider all the possible variants of equation (98) which satisfy this relation.

a) If all the three constants have the same sign, then equation (98) assumes the form

$$\rho = \rho_1 (c_1^2 + c_2^2) + \rho_3 c_3^2. \tag{100}$$

129

When $|\rho_3| > |\rho_1|$, this equation describes a surface which is similar to a prolate ovaloid of revolution, which is white for positive values of ρ_1, ρ_2, ρ_3 (Fig. 118) and black for negative values.

When $|\rho_3| < |\rho_1|$, equation (100) describes a surface which is similar to an oblate ovaloid of revolution (Fig. 119). It can also be either white or black, depending on the sign of the radius vectors.

Both these surfaces, as well as the spheres described above, do not possess symmetry elements of the second kind. Their symmetry corresponds to the $\infty : 2$ group of a twisted cylinder which we have frequently met before.

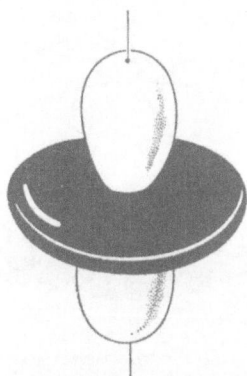

Fig. 120. Gyration surface of uniaxial crystals. Left-handed rotation along the vertical axis (positive); right-handed rotation along the horizontal axes (negative).

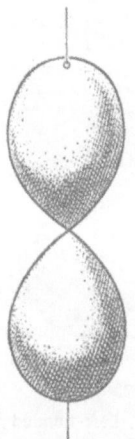

Fig. 121. Left-handed gyration surface of uniaxial crystals. Rotation along horizontal axes is absent.

b) If the constants ρ_1 and ρ_3 in equation (100) have different signs, then the equation can be written in the form

$$\rho = -\rho_1 (c_1^2 + c_2^2) + \rho_3 c_3^2. \tag{101}$$

The corresponding surface of revolution again has the $\infty : 2$ symmetry and is shown in Fig. 120. As can be seen from this figure, the surface consists of two white ovoid regions and one black torroidal region. The antipode of this surface is a surface in which the ovoid regions are black and the torroidal region is white. The equation of this surface can be obtained from equation (101) by interchanging the signs of ρ_1 and ρ_3.

c) If one of the constants (ρ_1) in the general equation (100) for gyration surfaces of uniaxial crystals is zero, then this equation assumes the form

$$\rho = \rho_3 c_3^2. \qquad (102)$$

The corresponding surface has the form of two ovoid regions in contact with each other (Fig. 121), and can be either white or black, depending on the sign of ρ_3. The symmetry of this surface is the same as the symmetry of all the other gyration surface of uniaxial crystals described so far, namely, $\infty : 2$.

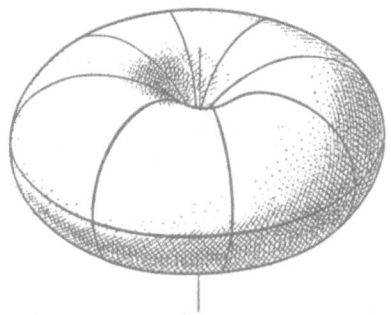

Fig. 122. Left-handed gyration surface of uniaxial crystals. No rotation along the vertical axis.

d) If ρ_3 is equal to zero then equation (100) assumes the form

$$\rho = \rho_1 (c_1^2 + c_2^2). \qquad (103)$$

The corresponding toroidal surface (Fig. 122) can be either white or black depending on the sign of ρ_1. In both cases it has the $\infty : 2$ symmetry.

e) When $\rho_1 = -\rho_{2'}$, $\rho_3 = 0$, equation (98) assumes the form

$$\rho = \rho_1 (c_1^2 - c_2^2). \qquad (104)$$

The corresponding surface (Fig. 123) consists of four, equal in magnitude and form, ovoid regions, two of which are white and the other black. It is easy to verify that this surface has the $\bar{4} \cdot m$ symmetry (Fig. 124), since the white (left) region is transformed into a black (right) region by reflection in the planes of symmetry and rotation through 90° about the mirror axis $\bar{4}$, while a white (black) region is transformed into a white (black) region by a 180° rotation about the double axes and a 180° rotation about the axis $\bar{4}$.

131

The important property of this surface is that, in distinction to all the above gyration surfaces, it has symmetry elements of the second

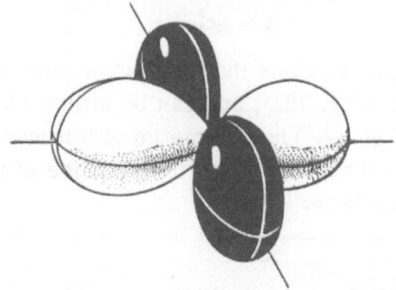

Fig. 123. Gyration surface of uniaxial crystals. Rotation along one horizontal axis is antiequal to the rotation along the other horizontal axis. There is no rotation along the vertical axis.

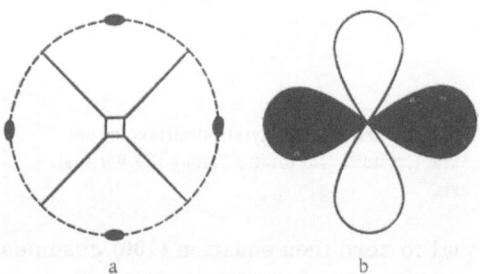

a
b

Fig. 124. Symmetry of the gyration surface shown in Fig. 123. a) Symmetry elements; b) the corresponding section of the surface.

kind, namely, a mirror axis $\bar{4}$ and two symmetry planes m. Similarly to any surface having symmetry elements of the second kind, the above surface does not possess a mirror antipode different from itself.

3. In biaxial crystals, the three constants are all different in absolute magnitude, i.e.,

$$|\rho_1| \neq |\rho_2| \neq |\rho_3|,$$

but their signs can be either different or not.

a) When the signs of the three constants are the same, equation (98) may be written in the general form

$$\rho = \rho_1 c_1^2 + \rho_2 c_2^2 + \rho_2 c_3^2. \tag{105}$$

The corresponding gyration surface which has the $2 : 2$ symmetry is shown in Fig. 125. The surface can be either white or black, depending on the sign of the constants. It has only three twofold axes, but no other symmetry elements, in particular, no symmetry planes. Its form resembles a general ovaloid.

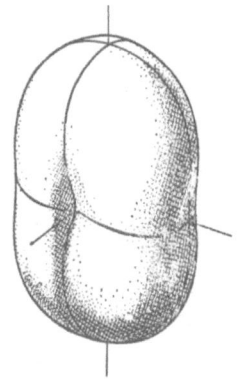

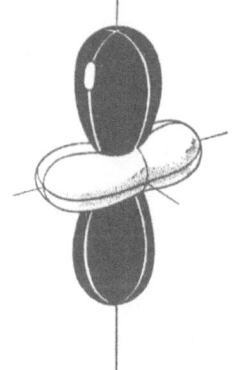

Fig. 125. Left-handed gyration surface of biaxial crystals. Left-handed (positive) rotation along the three symmetry axes.

Fig. 126. Gyration surface of biaxial crystals. Right-handed rotation along the vertical axis, left-handed rotation along the other axes.

b) If one of the three unequal constants, for example ρ_3, has a different sign than the other two, then equation (105) assumes the form

$$\rho = \rho_1 c_1^2 + \rho_2 c_2^2 - \rho_3 c_3^2. \tag{106}$$

The corresponding surface has the same symmetry as the previous one, namely, $2 : 2$ and is shown in Fig. 126. This surface consists (for negative ρ_3) of two black ovoid regions with vertices in contact, and a white region having the form of an oblong "boublik". When the signs of all the three constants in equation (106) are reversed, the resulting surface will have the same form as the previous one but the hues will be opposite to those given above.

c) If one of the constants, for example ρ_3, is zero, and the other two are unequal but have the same signs, then equation (105) assumes the form

$$\rho = \rho_1 c_1^2 + \rho_2 c_2^2. \tag{107}$$

The corresponding surface again has the $2 : 2$ symmetry and is shown in Fig. 127. It differs from the previous surface in that it has two

133

indentations whose vertices touch each other. The surface can be either white or black, depending on the sign of the two magnitudes of the specific rotation.

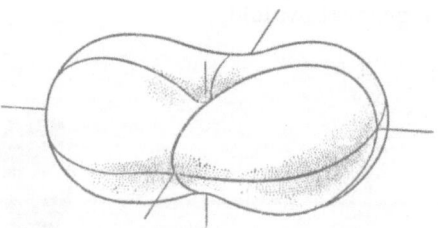

Fig. 127. Left-handed gyration surface of biaxial crystals. No rotation along the vertical axis.

d) If ρ_3 is zero and the other two constants are unequal and different in sign, then equation (105) assumes the form

$$\rho = -\rho_1 c_1^2 + \rho_2 c_2^2. \tag{108}$$

The corresponding surface, which again has the $2 : 2$ symmetry, is shown in Fig. 128. It consists of four ovoid regions which touch each

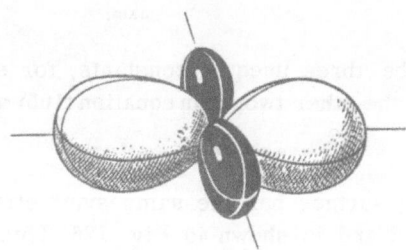

Fig. 128. Gyration surface of biaxial crystals. There is no rotation along the vertical axis while the rotation along the two horizontal axes is of opposite sign.

other at a point. Two of them are equal and white, and the other two are equal to each other, but not to the first two, and are black. When the signs of ρ_1 and ρ_2 are interchanged, the surface transforms into its antipode which has the opposite hues.

The above surfaces cover all the possible forms of gyration surfaces. They are summarized in Table VII.

134

TABLE VII

Symmetry and Equations of Gyration Surfaces

Symmetry	Equations
∞/∞	$\rho = \rho_1$
$\infty : 2$	$\rho = \rho_1 (c_1{}^2 + c_2{}^2) + \rho_3 c_3{}^2$ $\rho = -\rho_1 (c_1{}^2 + c_2{}^2) + \rho_3 c_3{}^2$ $\rho = \rho_3 c_3{}^2$ $\rho = \rho_1 (c_1{}^2 + c_2{}^2)$
$\overline{4} \cdot m$	$\rho = \rho_1 (c_1{}^2 - c_2{}^2)$
$2 : 2$	$\rho = \rho_1 c_1{}^2 + \rho_2 c_2{}^2 + \rho_3 c_3{}^2$ $\rho = \rho_1 c_1{}^2 + \rho_2 c_2{}^2 - \rho_3 c_3{}^2$ $\rho = \rho_1 c_1{}^2 + \rho_2 c_2{}^2$ $\rho = -\rho_1 c_1{}^2 + \rho_2 c_2{}^2$

Division of Crystals into Four Types According to the Symmetry of the Gyration Surfaces. Frequent reference was made in the above discussion to the important crystallophysical law, which in its exact formulation states that the morphological symmetry group of a crystal should either coincide with the nearest higher group of the property under consideration, or should be a subgroup of this group. Starting from this law, it is not difficult to pick out from the 32 groups of morphological symmetry those which are subordinate to each of the four groups of symmetry of the gyration surfaces. In this way we find, for example, that the groups subordinate to $2 : 2$ are the group $2 : 2$ itself and groups 1 and 2, groups subordinate to $\overline{4} \cdot m$ are the group $\overline{4} \cdot m$ itself and the groups $\overline{4}$, $2 \cdot m$, m, etc. A full discussion shows that among the 32 groups of morphological symmetry (classes of crystals) there are only 15 symmetry groups which are subordinate to the above four symmetry groups of gyration surfaces (Table VIII). As was to be expected, they include all the 11 classes of purely axial symmetry. It was also to be expected that the above Table will not include the 11 centrosymmetric classes, since inversion transforms any right-handed rotation ($-\rho$) into a coincident left-handed rotation ($+\rho$), which is only possible if ρ is zero. The important new result of the Chipart theory* described above is that rotation of the plane of

---
*This account of the theory is considerably different in form from the original version and the later interpretations given by Pockel and others.

TABLE VIII

Distribution of Morphological Symmetry Groups of Crystals Among
Four Symmetry Groups of Gyration Surfaces

Symmetry of gyration surfaces	$2:2$	$\bar{4}\cdot m$		$\infty:2$		∞/∞
Morphological symmetry of crystals	$2:2$	$\bar{4}\cdot m$	$2\cdot m$	$6:2$	6	$3/4$
	2	$\bar{4}$	m	$4:2$	4	$3/2$
	1			$3:2$	3	

polarization may (but need not necessarily) be found not only in crystals of purely axial symmetry, but in certain other crystals having symmetry elements of the second kind, and in particular, in crystals belonging to the groups m, $2\cdot m$, $\bar{4}$ and $\bar{4}\cdot m$. This result may appear to be unexpected since each separate ρ (a segment of a straight line with two circular arrows) does not possess symmetry elements of the second kind. It must, however, be borne in mind that the conclusions of the theory refer not to a single radius vector of a gyration surface but to all of them together, i.e., the gyration surface as a whole. In accordance with experimental data, the latter can have left-handed and right-handed radius vectors at the same time.

In conclusion, we note that the results which we have obtained for crystals can be extended to homogeneous media in general, and liquid crystals in particular. It is known that all liquid crystals, in the form in which they are usually studied, i.e., in unstressed state between the object and cover glasses, have the optical properties of uniaxial crystals. Liquid crystals which do not rotate the plane of polarization have the $m\cdot\infty:m$ optical symmetry, i.e., the symmetry of double refraction, which coincides with their morphological symmetry. Liquid crystals which do rotate the plane of polarization and, of course, show double refraction, have, in accordance with the general law of crystal physics, the lower symmetry $\infty:2$, which coincides with the symmetry of rotation of the plane of polarization and is subordinate to the $m\cdot\infty:m$ group.

Dispersion of Gyration Surfaces. It was pointed out above that the specific rotation of a crystal measured in a given direction depends on the frequency of the monochromatic light used. This means that the gyration surfaces of a given crystal should be somewhat different in form and position in space for monochromatic light of different color.

The dependence of the form and position of gyration surfaces on the frequency of light will be called the dispersion of gyration surfaces. This form of dispersion is completely analogous to the dispersion of indicatrices.

Just as in the case of the dispersion of indicatrices (p. 59), a single "monochromatic" gyration surface may differ in its symmetry from the family of all the monochromatic surfaces of the given crystal, and in fact, may turn out to be higher than the symmetry of the family of surfaces (according to the principle of Pierre Curie which states that the symmetry of a family of unequal figures is a combination only of the symmetry elements common to all the figures taken separately). One would therefore expect that the number of symmetry groups of such families of surfaces is greater than the number of symmetry groups of monochromatic planes, i.e., greater than four. It is not difficult to enumerate all these symmetry groups.

1. In triclinic crystals of class 1, the monochromatic gyration surfaces have a $2 : 2$ symmetry when taken separately, their axes are not parallel to each other and only the centers of the surfaces coincide. It follows that, in accordance with the Curie principle, none of these symmetry elements are found in the family of gyration surfaces (group 1).

2. In monoclinic crystals of class 2, all the monochromatic surfaces which have a $2 : 2$ symmetry when taken separately, have one of their 2 axes along the 2 axis of the crystal. The position of the other two axes of each monochromatic surface can be any of the possible positions. It is clear that the family of the gyration surfaces will have a 2 symmetry.

3. In monoclinic crystals of class m, monochromatic surfaces can only have a $\bar{4} \cdot m$ symmetry when taken separately, since there are no other symmetry groups containing m, among the symmetry groups of gyration surfaces. In crystals belonging to this class, all the monochromatic surfaces have one of their symmetry planes along the single m plane of the crystal. The position of the $\bar{4}$ axis and the 2 axes of the monochromatic gyration surfaces, can be any of the possible positions. As a result, the family of gyration surfaces has only this one plane of symmetry m.

4. In orthorhombic crystals of class $2 : 2$ isolated monochromatic surfaces also have a $2 : 2$ symmetry. The family of surfaces will clearly have the same $2 : 2$ symmetry.

5. In tetragonal crystals of class $\bar{4}$, the gyration surfaces have a $\bar{4} \cdot m$ symmetry when taken separately, their $\bar{4}$ axes are along the $\bar{4}$ axis of the crystal, while the m planes of the surfaces can take up any of the possible positions. As a result, the family of gyration surfaces will have the symmetry of a $\bar{4}$ crystal.

6. In tetragonal crystals of class $\bar{4} \cdot m$, and in orthorhombic crystals of class $2 \cdot m$, the gyration surfaces which separately have a $\bar{4} \cdot m$ symmetry have their $\bar{4}$ axis along the $\bar{4}$ axis, or the 2 axis, of the crystal, and their two m planes are along the two m planes of the crystal. As a result, the family of gyration surfaces will have the $\bar{4} \cdot m$ symmetry of the surfaces taken separately.

7. In uniaxial crystals of class $3, 4, 6, 3:2, 4:2, 6:2$, monochromatic gyration surfaces which have a $\infty : 2$ symmetry when taken separately, have their ∞-axes along the principal axis of the crystals. As a result, the family of gyration surfaces will have the same symmetry as each of the separate surfaces, namely, $\infty : 2$.

TABLE IX

Distribution of Morphological Symmetry Groups Among
Gyration Symmetry Groups

Gyration symmetry	1	2	$2:2$	m	$\bar{4}$	$\bar{4} \cdot m$		$\infty : 2$	∞/∞
Morphological symmetry of crystals	1	2	$2:2$	m	$\bar{4}$	$\bar{4} \cdot m$ $2 \cdot m$	$6:2$ $4:2$ $3:2$	6 4 3	$3/4$ $3/2$

8. In cubic crystals of class $3/4$ and $3/2$, the ∞/∞ symmetry of the family of gyration surfaces (spheres) will not differ from the symmetry of a single sphere.

It is clear from the above, that there are altogether eight groups of gyration symmetry of crystals, i.e., eight types of crystals which differ from each other in their ability to rotate the plane of polarization when they are studied in white light (when rotational dispersion is taken into account). The above distribution of morphological symmetry groups among the higher groups of gyration symmetry is summarized in Table IX. Again, the result obtained for crystals can be extended to homogeneous transparent media in general.

Examples of Optically Active Crystals. Before we pass on to the description of specific crystals which are able to rotate the plane of polarization, it is necessary to emphasize once more that optical activity is a possible but not a necessary property of crystals belonging to these fifteen groups of morphological symmetry which are subordinate to the eight groups of gyration symmetry. This means that, for example, if morphological criteria indicate that a given crystal belongs to the "quartz" group $3 : 2$, then one may expect that it will be optically active, while if morphological criteria indicate that the crystal belongs to some group which is not subordinate to gyration symmetry, for example the "calcite" group $\bar{6} \cdot m$, then it cannot be optically active under any circumstances. We shall discuss a number of examples of optically active crystals belonging to different systems.

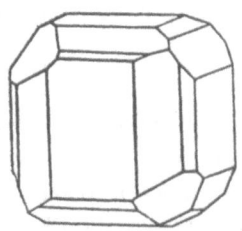

Fig. 129. Ideal form of crystals of sodium chlorate. Example of optically active cubic crystal.

1. Crystals of sodium chlorate ($NaClO_3$) belong to optically active cubic crystals. This substance crystallizes in the $3/2$ class from a water solution. The water solution itself is not optically active. Active right-handed and left-handed crystals (Fig. 129) separate out of the solution in equal quantities, in full agreement with the Pasteur rule. This means that the optical activity of these crystals is due to the dissymmetry of the structure and not the dissymmetry of the structural units, which in this case are the Na^+ and ClO_3^- ions which do not show any dissymmetry. The specific rotation of $NaClO_3$ for sodium light is $3.16°/mm$ at room temperature. On heating by $1°C$, the specific rotation increases by $0.00062°/mm$.

2. As we already know, quartz (which crystallizes in the $3 : 2$ class) is one of the optically active crystals in the hexagonal system. The following must be added to our discussion of quartz. Its gyration surface is of the form shown in Fig. 120. It follows that quartz is capable of rotating the plane of polarization not only along the optic axis, as we assumed earlier, but also in the direction perpendicular to the optic axis. The signs of the two rotations are opposite.

Let φ be the angle between a radius vector of the gyration surface and the optic axis. We then have $c_3 = \cos\varphi$, $c_2 = \sin\varphi$ and $c_1 = 0$ for radius vectors in one of the principal sections of quartz ($X_2 X_3$ plane).

139

Substituting these values into equation (101) we have

$$\rho = \rho_1 \sin^2 \varphi - \rho_3 \cos^2 \varphi. \qquad (109)$$

It has been established experimentally that the ratio $\frac{\rho_1}{\rho_2}$ remains constant for wavelengths between 546 mμ and 265 mμ and is equal to 0.54, i.e., the rotation of the plane of polarization in a direction perpendicular to the optic axis is lower by a factor of 2 compared with the rotation in the direction of the optic axis.

If the ratio $\frac{\rho_1}{\rho_2}$ is known, then the angle φ for the radius vector along which the rotation is absent ($\rho = 0$) can easily be determined from the latter formula. This angle is equal to 56°. A quartz plate cut along a plane perpendicular to this direction should behave in polarized light like an ordinary optically nonactive, doubly refracting plate. This means, in particular, that a linearly polarized ray with vibrations in the principal section should be transmitted by quartz in this direction without any changes in the character of its polarization.

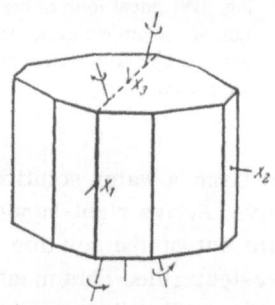

Fig. 130. Ideal form of Rochelle salt (potassium sodium tartrate). The optic axial plane is perpendicular to the X_2 axis. Rotation is right-handed along both optic axes. An example of an optically active orthorhombic crystal.

Red crystals of cinnabar (HgS), which have a particularly large specific rotation along the optic axis, also belong to optically active hexagonal crystals of the "quartz class" $3 : 2$. For the red rays transmitted by this crystal $\pm\rho_3 = 325°/mm$.

Among optically active crystals of the hexagonal system there are also crystals of potassium lithium sulfate which belong to class 6. The specific rotation of these crystals along the optic axis is $\pm 3°26'$ per millimeter (for sodium light).

3. Diacetyl phenolphthalein $C_{20}H_{12}O_4$ $(C_2H_3O)_2$ crystals which belong to the class $4 : 2$ are among the optically active crystals of the tetragonal system. Their specific rotation along the principal axis is 19.7°/mm for sodium light.

4. Among substances which crystallize in the orthorhombic system, Rochelle salt (potassium sodium tartrate) $KNa(C_4H_4O_6) \cdot 4H_2O$ (Fig. 130) is particularly noticeable. A water solution of technical

140

Rochelle salt precipitated from natural organic substances which are vinicultural products, rotate the plane of polarization to the right (synthetically, one obtains a ceramic mixture of levo and dextro salts, from which purely levo salt can be separated out). The optic axes of

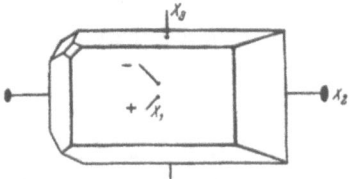

Fig. 131. Ideal form of sugar crystals. The optic axial plane is perpendicular to the X_2 axis. The rotation along the optic axes is of opposite sign.

crystalline Rochelle salt lie in the plane parallel to the (010) face, while in crystals of the ammonium analog of Rochelle salt $(NH_4)Na-(C_4H_4O_6) \cdot 4H_2O$, they are in a plane parallel to (100). In accordance with the $2 : 2$ symmetry, the two optic axes of Rochelle salt are equivalent to each other and hence the rotation along these axes is equal in magnitude and sign ($\rho_{Na} = -1.35°/mm$). The ammonium salt has an opposite sign of the rotation along the two optic axes ($\rho_{Na} = +1.55°$ per mm).

5. Crystals of the sugar $C_{12}H_{22}O_{11}$ and tartaric acid crystals mentioned earlier (Fig. 116), are among the optically active crystals of the monoclinic system. Sugar crystallizes in class 2 (Fig. 131). The optic axial plane in sugar crystals is perpendicular to the 2 symmetry axis. One of the optic axes is almost perpendicular to the cleavage plane (100). The other axis is at an angle $2V = 45°50'$ to the positive end of the X_3 axis, in sodium light. Sugar crystals have a left-handed rotation about the first axis ($\rho = +22°/mm$) and a right-handed rotation about the second axis ($\rho = -6.4°/mm$), in sodium light. In solution, sugar exhibits a right-handed rotation. The difference in magnitude and sign of the rotation along the two optic axis is in full agreement with the character of the symmetry of the crystal.

Tartaric acid crystals also belong to class 2. Technical tartaric acid in water solution is dextrorotary and is therefore known as dextro-tartaric acid. However, crystals of such d-tartaric acid exhibit left-handed rotation along both the optic axes which, in distinction to the axes in sugar crystals, lie in some irrational (i.e., not

141

having definite integral indices) plane parallel to the 2 axis. The optic axial angle for sodium light is $2V = 78°20'$. The acute bisectrix lies in the acute angle between the crystallographic axes $X\bar{Z}$. In accordance with the 2 symmetry, the optic axes are identical, and hence the rotation along them is the same ($\rho_{Na} = +11.4°$ per mm).

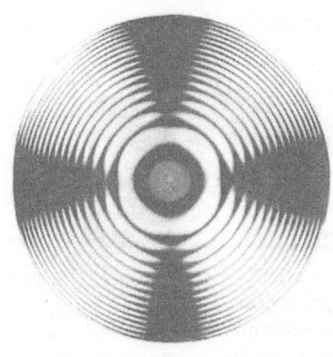

6. Crystals of organic substances, which have so far not been extensively studied, belong to the optically active crystals of the triclinic system. An example of such crystals is d-ethylenediaminecobaltichlorotartrate $C_4H_4O_6$-ClCo(N$_2$H$_4$C$_2$H$_4$)$_3$.

Fig. 132. Interference pattern obtained with a quartz plate cut perpendicular to the optic axis and observed in convergent light with crossed Nicol prisms. The black arms of the cross do not reach the center. The central part of the pattern is colored.

7. In addition to crystals and ordinary liquids, many anisotropic liquids (liquid crystals) having a $\infty : 2$ symmetry are also optically active. Some of them have enormous specific rotations, for example, cyanbenzalamine cinnamate, has the specific rotation $\rho = +11000°/mm$.

Observation of Rotation of the Plane of Polarization in Convergent Light. Optical activity of crystals can only be observed in its pure form along the optic axes. Along other directions it is masked by double refraction which rapidly increases as the angle between the rays and the optic axes increases. When one examines a plate cut from an active crystal, in the direction perpendicular to the optic axis in convergent polarized light between crossed Nicol prisms, one observes a pattern which differs from the corresponding pattern in the case of nonactive crystals only in that at the point corresponding to the exit of the optic axis there is no extinction, i.e., the arms of the dark cross of uniaxial crystals and the dark arm of biaxial crystals do not extend to the point where the optic axis intersects the pattern. This is illustrated in Fig. 132 which shows the interference pattern for quartz. Clearly, the interference hue observed in the middle of the field view in convergent white light will be the same as that observed in parallel white light. This hue will change when one of the Nicol prisms is rotated. In monochromatic light, the central colored region of the interference pattern can be extinguished by rotating one of the Nicol prisms through a suitable angle. Airy (1831) spirals are obtained in convergent polarized light when a $^1/_4 \lambda$ mica plate cut in the direction per-

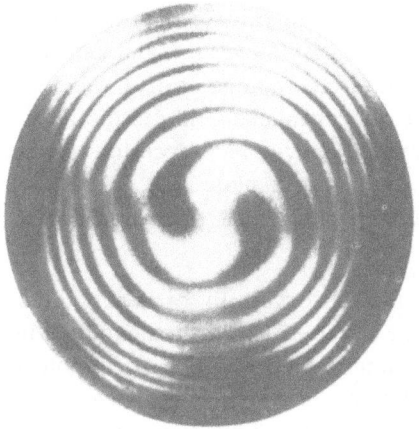

Fig. 133. Interference pattern obtained with a right-handed quartz plate and a superimposed mica plate, observed in convergent light with crossed Nicol prisms. The quartz plate is cut perpendicular to the optic axis. The Airy spirals are right-handed.

pendicular to the optic axis is placed on the quartz plate (Fig. 133). If the quartz plate is right-handed then the spirals are also right-

Fig. 134. Airy spirals. Right-handed quartz plate below the left-handed one. Convergent light, Nicol prisms crossed, quartz plates cut perpendicular to the optic axis.

Fig. 135. Airy spirals. Left-handed quartz plate below the right-handed one.

143

handed, and vice versa. If the mica plate is placed below the quartz plate the spirals bend in the opposite direction. Analogous effects are also observed in the biaxial case.

Another form of Airy spirals is obtained if two quartz plates of opposite sign but the same thickness, and cut in a direction perpendicular to the optic axis, are placed on each other (Figs. 134 and 135). These spirals have four branches, in distinction to the two branches of the spirals described above. The direction in which the spirals are bent depends on which quartz plate lies in the lower position. If the lower plate is right-handed then the spirals bend in the clockwise direction.

These effects are observed in crystal plates which are much thicker (by a factor of $10 - 100$) than the plates used to observe the interference phenomena which are due to simple double refraction. This is explained by the fact that in the case of rotation of the plane of polarization, i.e., circular double refraction, the quantity $n' - n''$ is much smaller than in ordinary double refraction.

From circular double refraction we naturally pass on to elliptical double refraction which is treated in the next chapter.

ELLIPTICAL POLARIZATION AND ELLIPTICAL DOUBLE REFRACTION. CONICAL REFRACTION

Elliptical Vibrations. If two vectors x_1 and x_2 execute harmonic vibrations along mutually perpendicular directions with the same frequency but different amplitudes A_1 and A_2 and different phases φ_1 and φ_2, then the resultant vector will, in general, execute elliptical vibrations, i.e., the locus of its end point will be an ellipse. This can be shown as follows.

Let the vibrations of the first vector be described by

$$x_1 = A_1 \sin \varphi_1 = A_1 \sin \omega t = A_1 \sin \tau,$$

then the vibrations of the second vector can, in general, be described by

$$x_2 = A_2 \sin \varphi_2 = A_2 \sin (\omega t - \Delta) = A_2 \sin (\tau - \Delta),$$

where $\Delta = \varphi_1 - \varphi_2$ is a constant which we shall call the phase difference [*].

Since in the study of elliptical vibrations the only important property is their form, i.e., the ratio of the semiaxes of the ellipses, their orientation in space and the character of the motion of the point describing the ellipse (left-handedness or right-handedness of the ellipse) and not the absolute dimensions, it follows that we may put $A_1 = 1$ or $\frac{A_2}{A_1} = k$. The latter quantity can conveniently be represented by the tangent of some angle α (the geometrical meaning of which will be described below), i.e., it is convenient to put $k = \mathrm{tg}\,\alpha$. Bearing in mind the above remarks, the vibrations of the two vectors may be described by

---

[*] The concept of phase difference must, in all cases, be carefully defined since it can mean different things under different conditions. For example, care must be taken to distinguish between the case when the harmonic vibrations of the two vectors are both sinusoidal and the case where one is sinusoidal and the other cosinusoidal.

145

$$x_1 = \sin \tau,$$
$$x_2 = k \sin (\tau - \Delta).$$
$$(110)$$

Let us rewrite these equations in the form

$$x_1 = \sin \tau,$$

$$x_2 = k (\sin \tau \cos \Delta - \cos \tau \sin\Delta).$$

Using the substitution $\cos \tau = \sqrt{1 - \sin^2\tau}$, we eliminate $\sin \tau$ and $\cos \tau$ from the above equation. As a result, we obtain the expression

$$k^2 x_1^2 + x_2^2 - 2kx_1 x_2 \cos \Delta = k^2 \sin^2 \Delta, \qquad (111)$$

which is an equation of the second degree in x_1 and x_2. In general, this equation describes an ellipse whose principal axes are at an angle to the X_1 and X_2 coordinate axes, but in special cases it may describe (1) an ellipse with principal axes along the coordinate axes, (2) a circle, and (3) a straight line.

1. When $\Delta = \dfrac{\pi}{2}$, $\dfrac{3\pi}{2}$, . . . equation (111) is transformed into the canonical equation of an ellipse, i.e.,

$$\frac{x_1^2}{1} + \frac{x_2^2}{k^2} = 1 \qquad (112)$$

with semiaxes $a_1 = 1$ and $a_2 = k$ directed along the coordinate axes.

2. When $\Delta = \dfrac{\pi}{2}, \dfrac{3\pi}{2}, \ldots$ and $u\,k = 1$ (i.e., $A_1 = A_2$), equation (111) becomes an equation of a circle, i.e.,

$$x_1^2 + x_2^2 = 1.$$

3. When $\Delta = n\pi$, the ellipse degenerates to a straight line given by

$$kx_1 = \pm x_2,$$

which, for even values of n passes through the origin at an angle α to the X_1 axis in the first quadrant, and for odd values of n at an angle $\pi - \alpha$ in the second quadrant. The geometrical meaning of the angle α thus becomes clear.

Equation (111) does not, either explicitly or implicitly, contain the time t . It follows that we cannot use this equation to determine the direction of motion of the point along the ellipse, i.e., we cannot determine whether the ellipse is a right-handed or a left-handed one. To do this, we must use the parametric form of the equation for the ellipse given by equation (110). In order to determine from these equations the character of the motion of the point describing the ellipse, it is sufficient to consider two points on the ellipse for two sufficiently close to each others values of t. In this way, it is easy to establish that for $0 < \Delta < \pi$ and at the same time, $k > 0$, the motion will be left-handed, and for $\pi < \Delta < 2\pi$ and $k > 0$, the motion is right-handed. Next, it can be easily verified that the character of the motion is preserved when Δ is increased or decreased by 2π, while a change in the sign of k or a change of $\pm \pi$ in Δ leads to a change in the character of the motion.

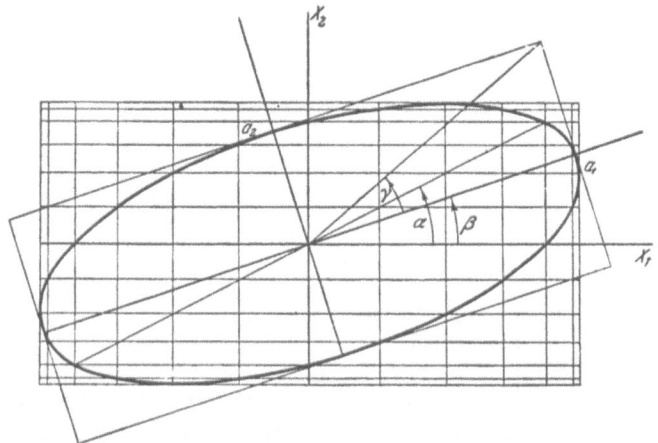

Fig. 136. Characteristic constants of an ellipse: semiaxes a_1, a_2 and angles α, β, γ.

The general equation of an ellipse (111) can always be reduced to the canonical form

$$\frac{x_1'^2}{a_1^2} + \frac{x_2'^2}{a_2^2} = 1$$

by rotating the coordinate system through an angle β which is the angle between the X_1 axis and the nearest principal axis (a_1 or a_2) of the ellipse (Fig. 136). We shall omit the proof and simply state that if α and Δ are known, this angle can be calculated from the formula

$$\operatorname{tg} 2\beta = \operatorname{tg} 2\alpha \cdot \cos \Delta. \tag{113}$$

In the description of elliptical vibrations, the ratio of the two principal semiaxes of the ellipse is often used and expressed as the tangent of some angle γ so that

$$\frac{a_2}{a_1} = \operatorname{tg} \gamma. \tag{114}$$

147

This angle γ can be calculated, using the expression

$$\sin 2\gamma = \sin 2\alpha \sin \Delta, \qquad (115)$$

which we quote without proof, provided α and Δ are known.

Ellipses corresponding to different values of Δ may be drawn, using the special coordinate net as shown in Fig. 137. These ellipses,

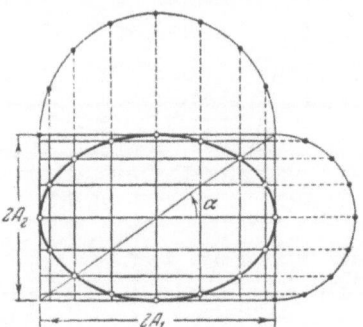

Fig. 137. Coordinate net used to construct ellipses.

in their most typical orientations, are shown in Fig. 138. In crystal optics, the important case is that of elliptical vibrations with $A_1 = A_2$ (i.e., with $k = 1$). These are shown in Fig. 139. The principal difference between elliptical vibrations and circular vibrations is that in elliptical vibrations the point moves nonuniformly over the ellipse while in the case of circular vibrations the motion is uniform.

Elliptical Polarization of Light Waves Due to Double Refraction in a Crystal Plate. Consider a linearly polarized monochromatic beam of light, incident normally (along the X_3 axis) on a crystal plate, in which the vibrations are described by

$$a = A \sin \tau$$

(Fig. 140). On entering the plate the rays are decomposed along the mutually perpendicular principal axes X_1 and X_2 of the section of the indicatrix, the corresponding vibrations being given by

$$\left.\begin{array}{l} x_1 = A \cos \alpha \sin \tau = A_1 \sin \tau, \\ x_2 = A \sin \alpha \sin \tau = A_2 \sin \tau, \end{array}\right\} \qquad (116)$$

where α is the angle between the direction of the original vibrations and the X_1 axis. The vibrations x_1 and x_2 have the same frequency and phase but different amplitudes, A_1 and A_2.

148

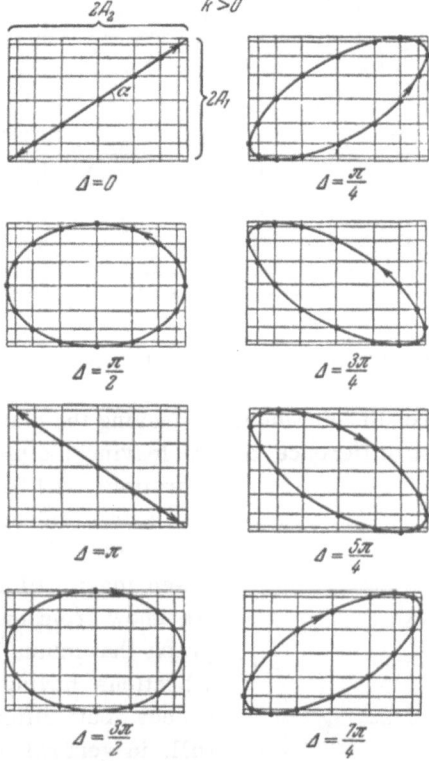

Fig. 138. Various forms of elliptical vibrations, which arise when natural light is passed through a Nicol prism and a crystal plate, when the angle α is kept constant and the phase difference Δ is variable and depends on the thickness of the plate. $A_1/A_2 = \operatorname{tg}\alpha = k$.

Inside the crystal plate, the x_1 and x_2 vibrations are propagated with different velocities (we are concerned with the normal velocities

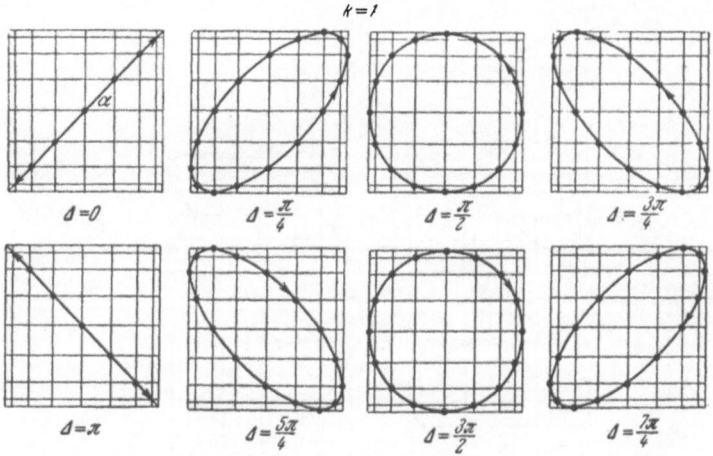

$k=1$

Fig. 139. Various forms of elliptical vibrations for $A_1=A_2$, i.e., for $\alpha=45°$.

v_N here) as a result of which, on leaving the plate, the vibrations will have a phase difference Δ. On leaving the plate, the x_1 and x_2 vibrations will be propagated with the same velocity in air, and hence the phase difference between them will remain constant. We know from the previous paragraph that mutually perpendicular vibrations having the same frequency but different amplitudes will, in general, combine to produce elliptical vibrations if there is a phase difference Δ between them. This means that the light emerging from the above crystal plate should be elliptically polarized, i.e., the light vector should execute elliptical vibrations.

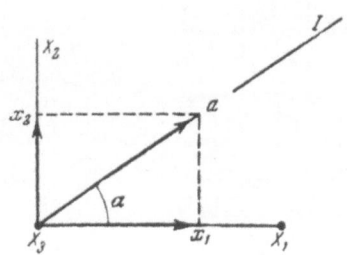

Fig. 140. Drawing illustrating the appearance of elliptical polarization in a crystal plate.

Dependence of the Form of Elliptical Vibrations on the Thickness of the Doubly Refracting Crystal Plate. $^1/_4 \lambda$ *Plates.* Equations (110) show that the form of elliptical vibrations is completely determined by the two quantities $k = \dfrac{A_2}{A_1}$ and Δ (or two other equivalent quantities). If the orientation of the plate relative to the direction of

150

vibrations in the Nicol prism is kept fixed ($\alpha =$ const) and only the thickness of the plate is varied, the only variable quantity in (110) is Δ. Thus, a study of the dependence of the form of elliptical vibrations on the thickness of the doubly refracting plate may be reduced to a study of this dependence on Δ, which we have already carried out. As was pointed out above, an especially important case occurs when $k = 1$ (shown in Fig. 139), i.e., when the plate is in a diagonal position with respect to the direction of vibrations in the Nicol prism.

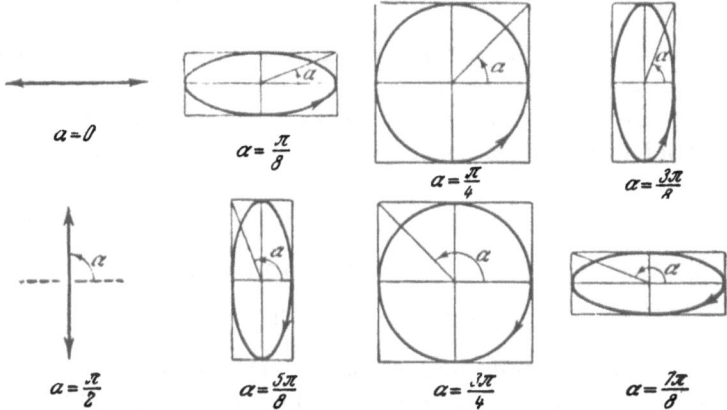

Fig. 141. Various forms of elliptical vibrations which are obtained when the plate is rotated about the vertical axis. The angle α is the variable.

In this case, by varying the thickness of the plate, it is possible to obtain all the forms of elliptical vibrations, including the limiting cases of linear and circular vibrations. The latter two special cases are obtained by putting $\Delta = \frac{\pi}{2}$, i.e., when the path difference produced by the plate is equal to a quarter of the wavelength of the incident monochromatic light ($\Gamma = \frac{1}{4} \lambda$). Such $\frac{1}{4} \lambda$ plates are often used to obtain circularly polarized Na light. These plates are usually made of mica. When such plates are observed in white light and between crossed Nicol prisms, they have a light grey interference hue of the first order. A combination of the first Nicol prism with a mica $\frac{1}{4} \lambda$ plate in a diagonal position is known as a circular polarizer, and a similar combination of a mica plate and the second Nicol prism, placed in the opposite order, is called a circular analyzer. A combination of the first Nicol prism with a mica plate such that the path difference in the diagonal position of the plate is neither zero nor $\frac{1}{4} \lambda$ is known as an elliptical polarizer, and a similar combination of a mica plate with the second Nicol prism is called an elliptical analyzer.

Changes in the Form of Elliptical Vibrations when the Plate is Rotated about Its Normal. All the forms of elliptical vibrations can be obtained using one quarter-wave plate by rotating the plate about its normal, when the latter coincides with the direction of the incident beam of light. In this case, the only variable quantity in (110) is $k = \frac{A_2}{A_1}$ which is equal to the tangent of the variable angle α. The general formula (111) for an ellipse now assumes (for $\Delta = \frac{\pi}{2}$) the canonical form (112) of the equation of an ellipse referred to its principal axes. This means that, in distinction to the elliptical vibrations described in the previous paragraph, we now deal with vibrations which are parallel to the direction of the linear vibrations within the plate.

On replacing the quantity k in (112) by $\frac{A_2}{A_1}$, and the quantities A_1 and A_2 by $A \cos\alpha$ and $A \sin\alpha$, we have, using (116),

$$x_1^2 \sin^2 \alpha + x_2^2 \cos^2 \alpha = \sin^2 \alpha. \tag{117}$$

In this form, the formula may be used to construct all the forms of vibrations corresponding to different values of α. It should be noted, however, that this construction can be carried out graphically without the use of the above formula (Fig. 141). A combination of a rotating Nicol prism and a $^1/_4$ λ plate is known as a Senarmont compensator.

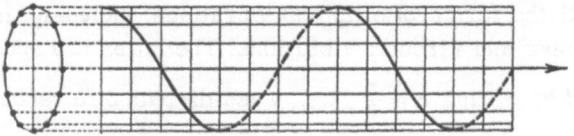

Fig. 142. Model of an elliptically polarized ray. An elliptical cylinder with a screw thread drawn on its surface. The curve moves uniformly in the direction of the axis of the cylinder.

Equation of Propagation of Elliptically Polarized Waves. From the mathematical point of view, elliptically polarized light may be looked upon either as a single beam with elliptical vibrations of the light vector, or as two interacting linearly polarized beams with vibrations x_1 and x_2. In purely computational problems, the second viewpoint is more convenient, although physically, the former appears to be the more correct. Using the second approach, the equation describing the propagation of elliptically polarized waves may be replaced by the following two equations of propagation of linearly polarized waves:

$$x_1 = A_1 \sin(\tau - \varkappa x_3),$$
$$x_2 = A_2 \sin(\tau - \varkappa x_3 - \Delta). \tag{118}$$

The first of these equations can be obtained from equation (29) by replacing ωt by τ and x by x_3. The two equations differ only in phase. The phase difference Δ is given by

$$\varphi_1 - \varphi_2 = (\tau - \varkappa x_3) - (\tau - \varkappa x_3 - \Delta) = \Delta.$$

Model of Elliptically Polarized Ray. Using equations (116), it is easy to calculate at a given instant of time ($t = \text{const}$) the magnitude and direction of the electric field for each point on the ray, and thus give an instantaneous picture of the distribution of these vectors along a ray. These calculations can, of course, be replaced by constructions based on the use of Fig. 136. It may be shown, using either of these methods, that the geometrical locus of the end points of the electrical vectors in an elliptically polarized ray is a spiral drawn on a cylinder (Fig. 142). It follows that such a spiral moving uniformly in the direction of its longitudinal axis may be used as a geometrical model of an elliptically polarized ray.

One notes the following similarities and differences between elliptically and circularly polarized rays.

Monochromatic circularly polarized light	Monochromatic elliptically polarized light
Light vectors are transverse with respect to the ray and rotate uniformly about the ray.	Light vectors are transverse with respect to the ray and rotate nonuniformly about it; their end points describe ellipses.
The end points of the light vectors form at each instant of time a screw thread on a circular cylinder.	The end points of the light vectors form at each instant of time a screw thread on an elliptical cylinder.
Circularly polarized light is completely determined by the magnitude of the light vector A,	Elliptically polarized light is completely determined by the amplitude A_1 and A_2 of the com-

Monochromatic circularly polarized light	Monochromatic elliptically polarized light
the wavelength λ and the velocity v.	ponent linearly polarized rays, the wavelength λ and the velocity v.
Circularly polarized rays differ in the sign of the rotation of the light vector relative to an observer looking against the ray.	Elliptically polarized rays differ in the sign of the rotation of the light vector relative to an observer looking against the ray.

Stationary Group Waves within a Doubly Refracting Plate.
The dependence of the form of elliptical vibrations on the thickness of a plate, as described above, can be represented by a model (Fig. 143) in which the thickness d is plotted along the vertical line and the form of the vibration is directly shown at different horizontal levels by the corresponding ellipses or, in special cases, circles and parts of straight lines. The envelope of these curves is a model of the special group waves formed as a result of superposition of two systems of waves linearly polarized in mutually perpendicular planes and propagated with different velocity for different wavelengths in the crystal plate. These group waves are stationary in the sense that their velocity $V = 0$. This follows from the method of constructing the model and also from formula (18):

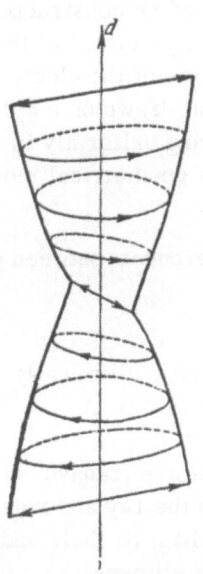

Fig. 143. Model showing the dependence of the form of vibrations on the thickness of the plate. The thickness is plotted on the vertical axis.

$$V = \frac{\lambda' v'' - \lambda'' v'}{\lambda' - \lambda''},$$

since in this case

$$\frac{\lambda'}{\lambda''} = \frac{v'}{v''}.$$

These waves divide the crystal plate into layers of equal thickness with periodically repeating forms of vibration. This thickness is equal to the group wavelength given by (16):

$$L = \frac{\lambda'\lambda''}{\lambda' - \lambda''} \ .$$

It must be noted that both here and in other similar cases, the stationary group waves ($V = 0$) do not remain unchanged in time since vibrations take place in them and are different at different depths in the crystal plate. In this sense our group waves are by no means stationary, similarly to the group waves discussed on p. 122.

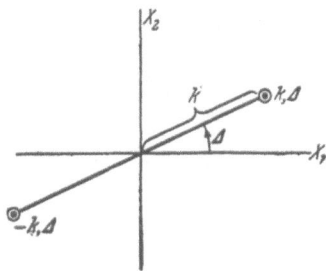

Fig. 144. The principle of the Poincaré diagram. A point whose coordinates are k, Δ represents elliptical vibrations characterized by the parameters k, Δ.

Poincaré Diagram (1892): Any elliptical vibration can, provided we are not interested in the absolute values of A_1 and A_2, be represented by the set of equations

$$\left. \begin{array}{l} x_1 = \sin \tau, \\ x_2 = k \sin (\tau - \Delta). \end{array} \right\} \tag{119}$$

Let us look upon Δ and k as polar coordinates on a plane (Fig. 144) and the axis from which Δ is measured as the X_1 axis of another, Cartesian set of coordinates (this axis is usually chosen so that it is parallel to the direction of vibrations in the crystal plate). Each point in the plane will then represent a vibration of a strictly defined form.

Let us consider which curves will, under these conditions, represent the most interesting and important families of vibrations.

1. If the characteristic points lie on the X_1 axis then $\Delta = 0$ and k is arbitrary. All the corresponding vibrations are linear. They include, for example, linearly polarized vibrations entering the crystal plate at various angles $\alpha = \text{arctg } k$ (Fig. 140).

When $k = 0$ the characteristic point is at the origin and the corresponding linear vibrations are parallel to the X_1 axis. In the diagram shown in Fig. 145 these vibrations are indicated by a heavy horizontal line (cf. Fig. 141).

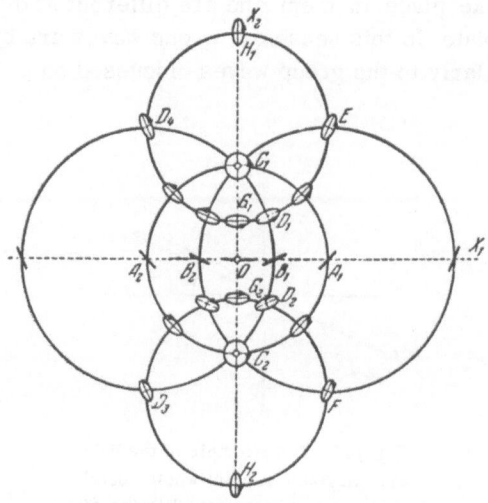

Fig. 145. The Poincaré diagram. The center of the diagram (i.e., the origin) represents harmonic vibrations along the X_1 axis. Each point on the X_1 represents harmonic vibrations at an angle α to this axis, and the angle increases as the distance from the center increases. The point A_1 represents vibrations at an angle of 45° to the X_1 axis. A point on the X_1 which is at an infinite distance from the center represents vibrations perpendicular to the X_1 axis. Each point on the X_2 axis represents elliptical vibrations with the principal axes of the ellipses parallel to the X_1, X_2 axes. All points on the X_2 axis which lie between C_1 and C_2 represent elliptical vibrations with major axes parallel to the X_1 axis; all other points on the X_2 axis represent elliptical vibrations with major axes parallel to the X_2 axis, and the points C_1 and C_2 themselves represent circular vibrations. All the points lying on a meridional circle (for example, B_1, D_1, C_2, D_2, D_3) represent elliptical vibrations with parallel principal axes. All the points on a circle drawn with the center on the X_2 axis (for example all the points on the circle D_1 E) represent elliptical vibrations which are such that the corresponding ellipses are identical except that they are rotated relative to each other. Points such as G_1, H_1 which lie at the ends of a diameter of this circle represent vibrations which are such that the corresponding elipses are equal but crossed. All points lying above the X_1 axis represent lefthanded elliptical vibrations, while all points lying below the X_1 axis represent lefthanded elliptical vibrations.

When $k = 1$ the characteristic point A_1 lies on the X_1 axis at a unit distance from the origin. The corresponding linear vibrations are at an angle of 45° to the X_1 axis and are shown in the diagram by a heavy line intersecting the X_1 axis at this angle.

When $k = \infty$ the characteristic point goes off to infinity along the X_1 axis. The corresponding linear vibrations are then perpendicular to the X_1 axis.

When $1 > k > 0$, for example at the point B_1 on the diagram, the linear vibrations are at angles $0 < \alpha < 45°$ to the X_1 axis.

Linear vibrations corresponding to the points A_2, B_2 are mirror images relative to the X_2 axis of vibrations corresponding to A_1, B_1.

2. If the characteristic points lie on the X_2 axis then $\Delta = \frac{\pi}{2}, \frac{3\pi}{2}, \ldots,$ and k can assume arbitrary values. It was explained earlier (p. 146) that under these conditions we have, in the general case, elliptical vibrations such that the principal axes of the ellipses are parallel to the X_1, X_2 axes.

In the special case $k = 0$ (point O at the origin) the linear vibrations are parallel to the X_1 axis.

Another special case is defined by $k = 1$ and $\Delta = \frac{\pi}{2}$ which gives left-handed circular vibrations, which are indicated by the point C_1 in the diagram and a small circle drawn about this point with an arrow indicating the left-handed nature of the vibrations.

In the general case, when $0 < k < 1$ and $\Delta = \frac{\pi}{2}$ we have left-handed elliptical vibrations with major axes along the X_1 axis (p. 147). In the diagram, these vibrations are indicated by points placed between O and C_1. One of these points is G_1 which is surrounded by a small left-handed ellipse oriented in the appropriate way.

When $\infty > k > 1$ and $\Delta = \frac{\pi}{2}$, the vibrations are executed over left-handed ellipses with major axes along the X_2 axis. In the diagram, these vibrations are indicated by points placed on the X_2 axis above the point C_1. One of these is the point H_1 which is surrounded by a small ellipse whose major axis points upward.

Left-handed vibrations represented by G_1, C_1, H_1, ..., can be associated with similar but right-handed vibrations indicated by G_2, C_2, H_2 ...

3. If the characteristic points lie on the circle $A_1C_1A_2C_2$, then $k = 1$ and Δ is arbitrary.

The corresponding family of vibrations was considered earlier (Fig. 139) and represents a set of ellipses (including the limiting cases of a circle and a straight line) whose principal axes are at angles $\alpha = 45°$ and $135°$ to the X_1, X_2 axes.

The two special points C_1, C_2 on the diagram, which represent circular vibrations, will be called the poles of the Poincaré diagram. The X_1 axis is then called the equator, all the circles passing through the points C_1, C_2 are called meridians, and circles orthogonal to them are called small circles of the diagram.

4. It may be shown that the principal axes of ellipses whose centers lie on any given meridian, for example $C_1D_1D_2C_2$, are parallel.

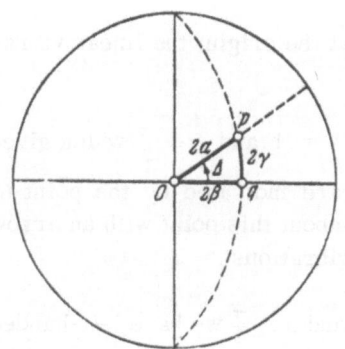

If two ellipses such as D_1, D_2 or D_3, D_4 both lie within the main circle $C_1A_1C_2A_2$, or outside this circle, their major and, correspondingly, minor axes will be parallel to each other, but if one of the two ellipses, for example D_2, lies within the main circle, and the other, say D_3, lies outside this circle, then the major axis of the one will be parallel to the minor axis of the other. This means that any meridian is a geometrical locus of points corresponding to elliptical vibrations with the same angle β.

Fig. 146. Diagram showing the relation between the quantities α, β, γ, Δ.

5. Similarly, it may be shown that all points which lie on a small circle correspond to elliptical vibrations with equal γ.

6. It is easy to verify that all points which lie above the X_1 axis correspond to left-handed vibrations, and all points below this axis to right-handed ones.

Poincaré Diagram as a Stereographic Projection of Points on a Sphere. The existence of two poles, an equator and circular meridians and parallels in the Poincaré diagram shows that this diagram is a stereographic projection of points on a sphere. We shall state without proof that any point P whose coordinates in the Poincaré diagram are k, Δ corresponds to a point p on a sphere with longitude 2β, latitude 2γ and angular distance of 2α from the origin O of these spherical coordinates, where α, β, γ are elliptical vibration parameters which were discussed earlier (Fig. 146).

These parameters are related by any three of the following equations

$$\left.\begin{aligned}
\cos 2\alpha &= \cos 2\gamma \cos 2\beta, \\
\mathrm{tg}\, 2\beta &= \mathrm{tg}\, 2\alpha \cos \Delta, \\
\sin 2\gamma &= \sin 2\alpha \sin \Delta, \\
\mathrm{tg}\, 2\gamma &= \mathrm{tg}\, \Delta \sin 2\beta.
\end{aligned}\right\} \tag{120}$$

They can be easily derived by reference to the rectangular spherical triangle Opq using well known formulas from spherical trigonometry. The second and third of these equations have already been quoted (cf. (113) and (115)).

The solution of various problems which may be reduced to the determination of the elliptical vibration parameters by means of a Poincaré diagram, can be considerably simplified if one uses the stereographic net due to Wulff, which gives directly the angular distances between points on a sphere and enables one to carry out operations on a plane which are equivalent to a rotation of the sphere about one of its diameters.

Combination of Elliptical Vibrations Using the Poincaré Diagram. Suppose we have two elliptical vibrations described by

$$x_1' = \sin \tau, \qquad\qquad x_1'' = \sin \tau,$$
$$x_2' = k' \sin (\tau - \Delta'). \qquad x_2'' = k'' \sin (\tau - \Delta'').$$

It is required to find the parameters k, Δ of the resultant vibration

$$x_1 = \sin \tau,$$
$$x_2 = k \sin (\tau - \Delta).$$

On adding the corresponding components we find that

$$x_1' + x_1'' = 2 \sin \tau,$$
$$x_2' + x_2'' = k' \sin (\tau - \Delta') + k'' \sin (\tau - \Delta'').$$

In order to reduce these expressions to the required form we divide both sides by 2. As a result we have

$$x_1 = \sin \tau,$$
$$x_2 = \frac{1}{2} [k' \sin (\tau - \Delta') + k'' \sin (\tau - \Delta'')] = k \sin (\tau - \Delta).$$

The latter equations are correct for arbitrary values of the variable τ. Putting $\tau = 0$ and $\tau = \frac{\pi}{2}$ we obtain from the second equation

$$k \sin \Delta = \frac{1}{2} (k' \sin \Delta' + k'' \sin \Delta''),$$

$$k \cos \Delta = \frac{1}{2} (k' \cos \Delta' + k'' \cos \Delta'').$$

On squaring and adding we find that

$$k = \frac{1}{2} \sqrt{k'^2 + k''^2 + 2k'k'' \cos (\Delta' - \Delta'') \ldots} \qquad (121)$$

On dividing the first equation by the second we obtain

$$\operatorname{tg} \Delta = \frac{k' \sin \Delta' + k'' \sin \Delta''}{k' \cos \Delta' + k'' \cos \Delta''}. \qquad (122)$$

This derivation has a simple geometrical interpretation illustrated in Fig. 147. The geometrical sum of the vectors k' and k'' which cor-

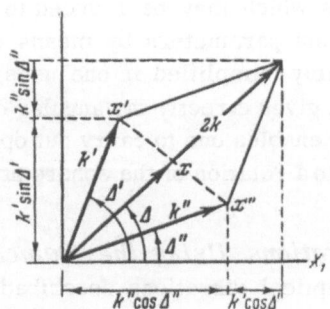

Fig. 147. Geometrical significance of the quantities which enter into formula (122).

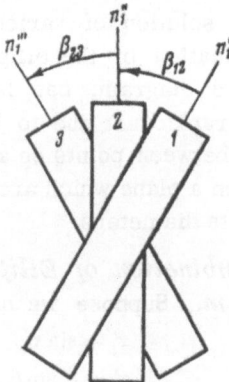

Fig. 148. Method of superposition of crystal plates.

respond to the phases Δ' and Δ'' is equal to the vector $2k$ which makes an angle Δ with the X_1 axis. It follows that the point representing the resultant vibration is the center of the parallelogram whose sides are k' and k''.

So far we have only considered the combination of elliptical vibrations; the decomposition of an elliptical vibration into two elliptical vibrations of a different form having given Δ', Δ'' (or k', Δ') can be carried out by a corresponding converse construction. It is important to note that the fact that such constructions are possible indicates that any elliptical vibration can always be formally represented as a sum of two, and consequently many, elliptical vibrations of other form.

Transmission of Linearly Polarized Rays through a Pile of Doubly Refracting Plates. Consider a pile of transparent doubly refracting crystal plates placed on top of each other in a fan-like arrangement (Fig. 148). Let us denote the direction of vibrations of the lower velocity waves propagated in plates 1, 2, 3, . . . by n_1', n_1'', n_1''', . . ., the angles between these directions by β_{12}, β_{23}, . . . and the phase differences due to the plates separately when monochromatic rays are

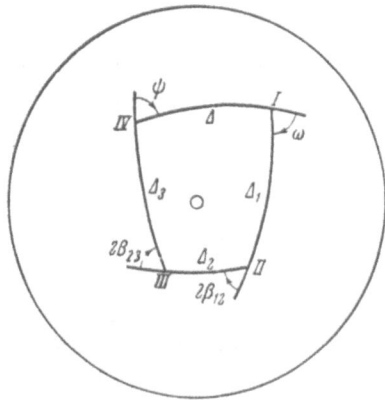

Fig. 149. Graphical determination of ρ and Δ when Δ_1, Δ_2, Δ_3, ... and the angles β_{12}, β_{23}, ... are given.

transmitted by them by Δ_1, Δ_2, Δ_3, . . . Using the Poincaré diagram it may be shown that in the case of linearly polarized light such a pile of plates is equivalent to a single active plate which rotates the plane of polarization through an angle ρ and one doubly refracting plate giving the required phase difference Δ. These quantities may be determined graphically in the following way by means of the Poincare diagram.

Using the Wulff net we draw, in stereographic projection, one of the great circles of a sphere with a radius $R = 1$ (Fig. 149). Next, at an arbitrary point I on this circle we mark off an arc equal to Δ_1. Starting from the point II on this arc we draw the next great circle of the sphere at an angle of $2\beta_{12}$ to the first one and mark off an arc equal to Δ_2 on it. From the end III of this arc we construct another great circle at an angle $2\beta_{23}$ to it and mark off an arc equal to Δ_3 on it. If the pile contains more than three plates, this construction is

161

continued until all the given quantities Δ_1, Δ_2, Δ_3, . . ., are used up, after which the last arc which closes the polygon $\Delta_1\Delta_2\Delta_3$ . . . is drawn. In the case of three plates this final arc is the arc I IV of a great circle. Calculations show that it is equal to the required path difference Δ and the area S of the polygon is twice the angle of rotation of the plane of polarization, i.e.,

$$2\rho = S. \tag{123}$$

Of course, if this area is expressed (for $R = 1$) in fractions of the surface area of the sphere (4π) then ρ will be in radians.

A number of interesting results follow from the above discussion.

1. It is well known that the result of a number of successive rotations about axes intersecting at a point in general depends on the order in which the rotations are carried out. This means that the effect of a pile of plates will be different if any two of the plates are interchanged.

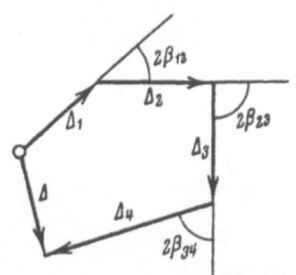

Fig. 150. Determination of the quantity Δ as the geometrical sum of the vectors Δ_1, Δ_2, Δ_3, ... in the case when the latter quantities are small.

2. Light transmitted through a pair of plates in which the vibrations are not parallel cannot be extinguished by rotating them between crossed Nicol prisms. In fact, for finite values of Δ_1, Δ_1, $2\beta_1$ the area of the spherical triangle with sides Δ_1, Δ_2, Δ and one of the external angles equal to $2\beta_{12}$ cannot be equal to zero. This means that a pair of doubly refracting plates is in the case of linearly polarized light always equivalent to one doubly refracting and one rotating plate. Suppose now that the polarized light from the first Nicol is first transmitted by the rotating plate and then by the double refracting plate. It is clear that the second plate has the same effect as when the Nicol prisms are not crossed. It is known, however, that in this case extinction cannot be obtained by rotating this plate (p. 77). It is easy to verify that this is also the case when the plates are placed in the reverse order.

3. Let us now consider the action of a pile consisting of a small number of very thin doubly refracting plates. Since in this case the quantities Δ_1, Δ_2, Δ_3, . . . are small, it follows that the corresponding sides of the spherical polygon are approximately straight lines and

the polygon itself may be looked upon as a flat polygon. It is known, however, that the area S of a plane polygon of given form is proportional to the square of one of its sides, i.e., $S \sim \Delta_1^2$. This means that when the Δ_1, Δ_2, Δ_3, . . . are small and there are not many of them, the area S, and consequently the rotation of the plane of polarization, may be neglected. It follows that a pile consisting of a small number of very thin plates is equivalent, in the case of linearly polarized light, to one doubly refracting plate producing a phase difference Δ . As shown in Fig. 150, the latter phase difference may be determined by measuring the length of the closing side of the polygon Δ_1, Δ_2, Δ_3, . . . in accordance with the vector addition law.

In the case of many, though very thin, plates, the area of the polygon $\Delta_1\Delta_2\Delta_3$ . . . may turn out to be appreciable, and with multiple rotational superposition of the polygon on itself, the area may become quite large. In that case it cannot, of course, be neglected.

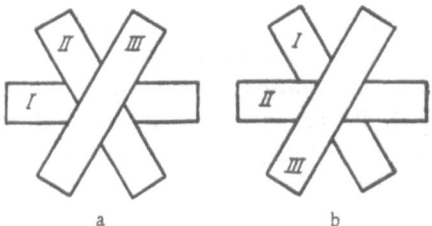

a   b

Fig. 151. Mica plate piles used by Reusch, with left-handed (a) and right-handed (b) screw arrangement. The effect of (a) on convergent polarized light is similar to the effect of left-handed quartz, and the effect of (b) is similar to that of right-handed quartz.

Moreover, when the plates are suitably superimposed, the quantity Δ may practically be reduced to zero, as a result of which the pile of doubly refracting plates will act not as one doubly refracting plate but as one optically active plate. This will be discussed in detail below.

Rotation of the Plane of Polarization Due to the Fan-like Arrangement of Doubly Refracting Plates. In 1869 Reusch showed that a pile of similar mica plates, in which the angle between successive plates is constant, has optical properties which are close to those of a single optically active plate. In particular, a pile consisting of three mica plates at 60° to each other, and arranged in a fan-like system, gives in convergent polarized light a result similar to that obtained by means of a quartz plate cut in the direction perpendicular to the optic

163

axis. A right-handed arrangement of the plates corresponds to right-handed quartz plate, and a left-handed arrangement corresponds to a left-handed quartz plate (Fig. 151).

The above method of studying elliptical polarization leads to a complete explanation of Reusch's observations. In fact, three mica plates superimposed on each other in a form of a right-handed fan can only be equivalent to a single rotating plate if in the spherical quadrilateral with sides Δ_1, Δ_2, Δ_3, Δ and external angles $2\beta_{12}$, $2\beta_{23}$, the fourth and closing side Δ, which characterizes the double refraction of the pile, is zero. In this case, however, we shall have not a quadrilateral but a triangle. In the case of the Reusch's mica pile, this spherical triangle should be equilateral and its external angles should be equal to 120°. However, such spherical triangles do not exist. An equilateral triangle with external angles equal to 120° can only be a plane rectilineal triangle. It follows that the mica pile used by Reusch can be replaced by one rotating plate to the same degree of approximation to which a spherical triangle can be replaced by a plane one. This means that the mica plates must be sufficiently thin. Naturally, the angle of rotation of the plane of polarization will, in this case, be small. If one wishes to increase this angle one has to use a sufficient number of such piles in tandem so that a sufficiently thick set of plates is obtained.

A quantitative estimate of the effect of such a set of plates can be carried out as follows.

The area of a plane equilateral triangle with sides $\Delta_1 = \Delta_2 = \Delta_3$, is given by

$$S = \frac{\Delta_1^2 \sqrt{3}}{2}.$$

We know already that the angle ρ_{3d} of rotation of the plane of polarization produced by one pile consisting of three plates of thickness d should be equal to one half of this area, i.e.,

$$\rho_{3d} = \frac{\Delta_1^2 \sqrt{3}}{4}.$$

Since according to (68)

$$\Delta_1 = \frac{2\pi d (n' - n'')}{\lambda},$$

it follows that

$$\rho_{3d} = \frac{\pi^2 d^2 (n' - n'')^2 \sqrt{3}}{\lambda^2}.$$

164

In order to find the specific rotation ρ, the angle ρ_{3d} should be divided by the thickness of the pile $3d$ so that

$$\rho = \frac{\pi^2 d \, (n' - n'')^2 \sqrt{3}}{3\lambda^2} \qquad (124)$$

Thus the specific rotation produced by a set of mica plates is inversely proportional to the square of the wavelength. It was shown earlier (cf. (93)) that the specific rotation produced by an optically active crystal is also inversely proportional to λ^2. It may, therefore, be concluded that the rotation of the plane of polarization in crystals (if it is not due to molecular asymmetry but asymmetry in the crystal structure) is produced by a fan-like disposition of the lattice layers. This is in fact the case in quartz where the atomic planes have a fan-like arrangement, just as in the mica piles used by Reusch.

Let us now consider in greater detail the conditions under which the pile of three or more doubly refracting plates can be exactly equivalent to a single rotating plate.

As was pointed out above, in the case of n plates, the spherical polygon with sides Δ_1, Δ_2, Δ_3, . . . and external angles $2\beta_{12}$, $2\beta_{23}$, . . . should be closed ($\Delta = 0$), i.e., it should be an n-sided polygon and not an $(n+1)$-sided polygon. When $\Delta_1 = \Delta_2 = \Delta_3 = . . .$, all the external angles of the n-sided polygon $2\beta_{12}$, $2\beta_{23}$, . . . should also be all equal. Their sum in a spherical polygon is always less than 2π. It follows that when $n = 3$, and $\beta_{12} = \beta_{23} = \beta_{31} < 60°$, i.e., a pile of three similar mica plates may be exactly equivalent to a single rotating plate only if the mica plates are superimposed in a fan-like fashion at an angle smaller than $60°$. The angle can always be calculated if Δ_1 is given, and it is by no means impossible to obtain angles β_{12} equal to an irrational fraction of the circle. For example, in the case of $1/8 \lambda$ plates, i.e., when $\Delta_1 = \frac{\pi}{4}$, the angle β_{12} should be equal to $57°14'$.

A special case of the fan-like arrangement of plates occurs when β_{12} is infinitely small and the number of plates in a complete fan is infinitely large. The corresponding structure should, clearly, have the symmetry of a double screw or a twisted ribbon. This symmetry is characterized by the presence of an infinite-fold screw axis of symmetry which is at the same time a twofold axis, and an infinite number of transverse twofold axes.

Such a symmetry can reasonably be ascribed to elastically sheared liquid crystal bodies. For example, it is known that p-azoxyphenetole in its normal unstressed liquid crystal state is optically inactive. However, this substance can be made to rotate the plane of polarization if it is placed in the form of a uniform layer between two microscope slides and one of the slides is rotated relative to the other about the normal. The angle of rotation of the plane of polarization should, if it is not greater than 90°, be then strictly equal to the angle through which the slide has been rotated. When the angle of rotation of the glass slide passes through 90° a rapid change in the sign of the rotation of the plane of polarization takes place. If the rotation of the glass slide is right-handed and is carried out through an angle $(90+n)°$ the effect is the same as when the specimen is rotated in the left-handed direction through an angle $n°$.

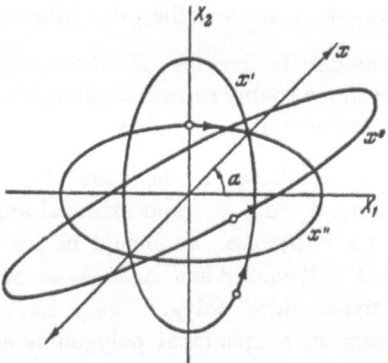

Fig. 152. Elliptical double refraction in the case where a linearly polarized ray enters the crystal plate at an angle $\alpha = 45°$ to the X_1 axis. The incident vibrations x are decomposed into x' and x'' within the plate. On leaving the plate, the vibrations recombine to form the elliptical vibrations $x°$.

General Discussion of Elliptical Double Refraction. In the foregoing chapters we discussed the phenomenon of double refraction of linearly and circularly polarized rays. It is natural to expect that the double refraction of elliptically polarized rays is also possible in crystals since linear and circular polarization are only special cases of elliptical polarization. Elliptical double refraction can be represented in the following way.

Suppose that the crystal plate is parallel to the plane of the paper (Fig. 152). It is assumed that there are two directions X_1 and X_2 in the plane of this plate which play the role of generalized vibration directions. If a monochromatic linearly polarized ray with vibrations x at an angle $\alpha = \frac{\pi}{4}$ to the X_1 axis is incident normally on the plate, then on entering the plate it will be split into two elliptically polarized rays. The corresponding ellipses x' and x'' are equal to each other and their principal axes are crossed. The latter axes coincide in direction with the X_1, X_2 axes. The motion of the end point of the electric vector in one of the ellipses (x') is in the counterclockwise direction, and in the other (x'') in the clockwise direction. The periods of the two elliptical vibrations are equal. The ratio of the axes of the ellipses depends on the way in which the crystal has been cut and also on the nature of the crystal itself. In the figure, the ratio of the major axis to the minor axis is taken to be equal to 2. The two rays are propagated with different velocities inside the crystal so that when they leave the crystal plate there is a different phase difference between them than when they entered the plate. Outside the crystal, the elliptical vibrations recombine, and as a result, the $x°$ vibrations are formed, which, in general, will also be elliptical. The corresponding ellipse will be inclined to the X_1, X_2 axes. It is clear from the drawing that in the example under consideration, we are dealing with right-handed elliptical double refraction since the major axis of the $x°$ ellipse is rotated in a clockwise direction relative to the vibration plane of the incident ray.

We assumed that the vibrations in the incident ray take place along the bisectrix of the angle between the X_1, X_2 axes. In a more general case, i.e., when $\alpha \neq \frac{\pi}{4}$, the linearly polarized ray will decompose into two elliptically polarized rays with vibrations over crossed ellipses, which will not be equal but only similar, i.e., the ratio of the axes of the two ellipses x', x'' will be the same and independent of α. The other properties are as before: the two rays are transmitted with different velocities through the crystal and on leaving the crystal they form a new, elliptically polarized ray $x°$.

If a natural unpolarized ray incident on a crystal plate may be looked upon as linearly polarized with a variable angle α, then at any given time this ray will be decomposed within the crystal into two elliptically polarized rays, as described above. As a result, we shall obtain, on the average, two elliptically polarized rays within the crystal, but these rays will no longer be coherent (cf. pp. 21 and 91).

Just as a linearly doubly refracting plate transmits linearly polarized rays without any change, provided the vibration direction in the rays is parallel to one of the vibration directions in the plate, so also an elliptically doubly refracting plate will transmit elliptically polarized rays without any change, provided the vibrations in these rays are similar and parallel to the vibrations in the plate.

On comparing linearly, circularly and elliptically doubly refracting plates we find the following differences between them. Linearly doubly refracting plates transmit without change linearly polarized rays four times during a complete revolution about the normal. Elliptically doubly refracting plate transmits without change the corresponding elliptically polarized rays twice during a complete rotation. When the plate is rotated through 90° from the position where it transmits elliptically polarized rays, the plate will completely extinguish the elliptical vibrations since the plate is now in the position where it transmits elliptical vibrations of the opposite sign. Circularly doubly refracting plates transmit circular vibrations for all angles of rotation about the normal. These differences between the three forms of vibrations are fully explained by the difference in their symmetry. Linear vibrations have a $2 \cdot m$ symmetry, elliptically polarized vibrations have a 2 symmetry, and circularly polarized vibrations have a ∞ symmetry.

Experimental Verification of the Existence of Elliptical Double Refraction. We began our exposition of optical crystallography with the description of linear double refraction in calcite. This phenomenon was confirmed by the directly observable decomposition of a beam of natural light into two beams, linearly polarized in two mutually perpendicular planes, and propagated along different directions. The experimental confirmation of the existence of circular double refraction was supplied by the Fresnel experiment, which showed that a beam of natural light may be decomposed by means of a quartz prism into two rays propagated along different directions and circularly polarized in opposite directions. Clearly, the existence of elliptical double refraction may be established by a similar experiment. Such an experiment was carried out by Croullebois in 1873.

In this experiment, the natural beam of light was passed through a quartz parallelepiped consisting of two triangular prisms as shown in Fig. 153. One of these prisms R was made from right-handed quartz and the other L from left-handed quartz. The refracting angles of the two prisms were made equal to 82°. The optic axis of quartz in the *R* prism was in the *ABCD* plane at an angle of 80° to the face *AB*. In the *L* prism the optic axis was in the *ADFE* plane, which was normal to the *ABCD* plane and made an angle of 80° with the *CDKI* face. A ray of unpolarized light was incident normally on the face

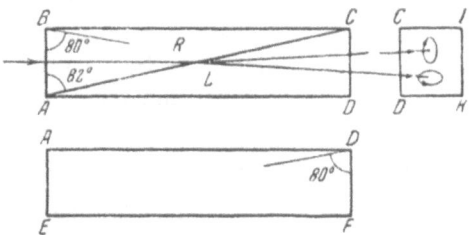

Fig. 153. Croullebois prism for the experimental confirmation of the existence of elliptical double refraction. The prism consists of two parts *R* and *L*, the former being made of right-handed, and the latter of left-handed quartz. The optic axis of *R* is at an angle of 80° to the face *AB* and lies in the upper plane *ABCD*. The optic axis of *L* lies in the side-face *ADFE* at the same angle to the rear face *CD)* (which is the same as **DF**).

AB of the *R* prism and thus made an angle of 10° with the optic axis of the *R* quartz prism. Under these conditions, the small circular double refraction in quartz should be a maximum and the large linear double refraction should be near its minimum. In other words, under these conditions the two effects should be of the same order of magnitude, and none of them should predominate and interfere with the observation of the other. Experiments showed that when the material used to cement the two faces together was suitably chosen (suitable refractive index), the incident ray could be separated into two rays polarized in crossed ellipses, in accordance with the requirements of the theory.

Conical Refraction. So far we have considered all the cases of double refraction in crystals, i.e., all cases of the division of the incident ray into two rays within the crystal plate. It remains to discuss one other special case in which a ray of natural light is split within the crystal not into two, but an infinite number of linearly polarized rays. This phenomenon was theoretically predicted by Hamilton in 1832 and was confirmed by Lloyd in 1833, and is now known as conical refraction.

169

If a very narrow parallel beam of natural light is incident normally on a plate cut in a plane perpendicular to one of its optic axes, then

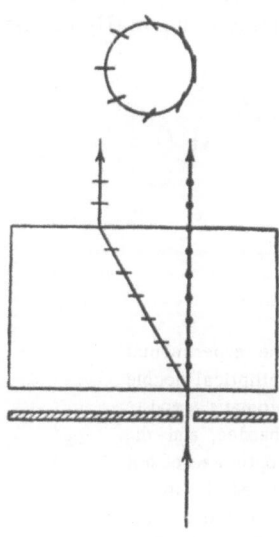

within the crystal, the beam is transformed into a hollow cone of rays, which on leaving the plate becomes a hollow cylinder. If a screen is placed perpendicular to this beam a luminous ring can be seen through a magnifying glass, and the diameter of this ring is independent of the distance of the screen from the plate (Fig. 154).

If this luminous ring is observed through a Nicol prism, it is found that it is not equally bright at all points and that the brightness minimum (darkness) lies at the point diametrically opposite to that of the brightness maximum. When the Nicol prism is rotated about a normal to the plate, the brightness minimum follows the rotation and traverses the entire ring. This shows that the rays forming the ring are linearly polarized in different azimuths as shown in the drawing.

Fig. 154. Observations of internal conical refraction in biaxial crystals. The incident ray passes through a fine hole in a screen and enters a thick crystal plate cut perpendicular to one of the optic axes. Inside the crystal, the ray is decomposed into a large number of other rays forming a conical surface. On leaving the plate, the rays form a cylindrical surface and produce a circular luminous image on a screen. The diagram shows the vibration directions at various points on the ring.

A qualitative interpretation of this phenomenon, which is known as internal conical refraction, can be obtained from a consideration of the section of the wave surface which includes the v_1, v_3 axes (Fig. 155). The NS plane, which is normal to the plane of the paper and tangential to the circle and the ellipse of the section of the wave surface, is also tangential to the wave surface at an infinite number of other points which are located on a circle above the wave surface "dimple". Rays drawn from the center O of the wave surface to all points on this circle form an elliptical cone SON. We know that each of these rays has, within the crystal, an associated wave normal, and conversely, each normal has an associated ray. We also know (Fig. 43) how to find the direction of the normal and the corresponding v_N velocity if the direction of the ray and the magnitude of the ray velocity v_S is known, and conversely, if the direction of the normal and the magnitude of the v_N velocity are known, we can find the direction of the associated ray and its ray velocity v_S. In the special

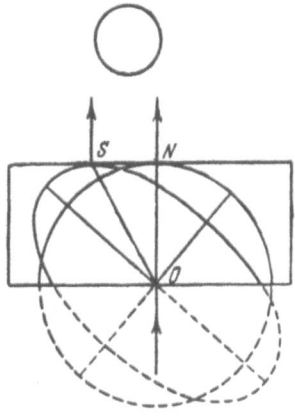

Fig. 155. Diagram illustrating the phenomenon of internal conical refraction.

case under consideration, if we carry out the construction for obtaining the associated ray when the wave normal is given, we find that a given normal has an infinite number of associated rays, and hence when the light is incident normally on to the plate, one will observe the separation of the incident ray into an infinite number of linearly polarized rays forming a cone. All these rays should, according to the law of refraction, emerge from the plate in the same direction in which they entered it, and hence the conical beam should be transformed into a cylindrical beam.

The above qualitative explanation of internal conical refraction can be treated analytically and it may be shown that the section NS of the conical beam is circular and the cone itself is, naturally, elliptical. As regards the distribution of vibration directions with respect to the azimuth, this distribution can be explained by the general rule for the determination of the vibration direction in a ray which was given earlier (Fig. 29), and it is found that this direction always lies in the NS plane and coincides with the normal to N.

In biaxial crystals, in addition to internal conical refraction one can also observe external conical refraction. For this purpose one uses a crystal plate cut along a plane perpendicular to a biradial

171

(Figs. 156 and 157). A convergent beam of natural light is incident on the plate so that the focus of the beam lies on the surface of the

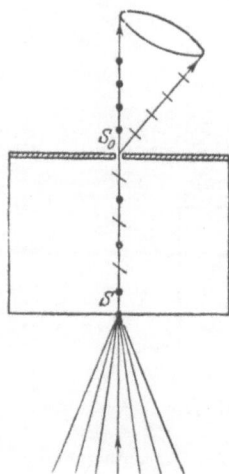

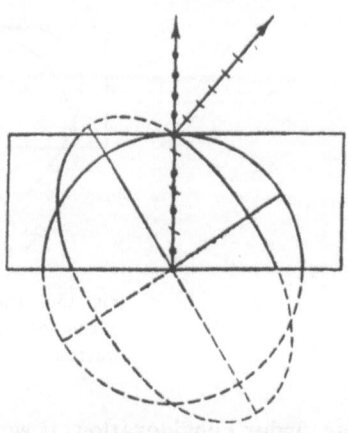

Fig. 156. Observation of external conical refraction in biaxial crystals. A conical beam of light is focused at the point S on the crystal plate cut perpendicular to one of the biradials of the crystal. The light leaves the plate through a fine hole S, in a screen placed as shown. The rays leaving the plate form a conical surface and give a circular luminous image on a screen, such that the diameter of the image increases as the distance between the screen and the plate increases.

Fig. 157. Diagram illustrating the phenomenon of external conical refraction.

plate. Such a ray contains all those rays which after refraction should travel along the biradial. Only these rays can emerge from the plate through the small aperture in the screen placed on the plate. On leaving the plate these rays will diverge and will give a luminous ring on a screen, the diameter of which will increase as the distance of the screen from the plate increases. The divergence of rays which travel in the same direction within the crystal can be explained by the fact that the associated normals have different refractive indices.

Conical refraction can in practice only be observed in a small number of crystals, namely, in those in which the angle SON (Fig. 155) is sufficiently large. However, even in these crystals a sufficiently thick plate must be used. Normally, one uses plates prepared from a crystal of rhombic sulfur in which the above angle is equal to 7°11'.

The angle α of the cone of rays in the case of internal conical refraction can always be calculated in advance from the formula

$$\operatorname{tg} \alpha = n_1 n_3 \sqrt{\left(\frac{1}{n_1^2} - \frac{1}{n_2^2}\right)\left(\frac{1}{n_3^2} - \frac{1}{n_2^2}\right)},$$

provided the principal refractive indices n_1, n_2, n_3 of the crystal are known.

DOUBLE ABSORPTION OF LIGHT IN CRYSTALS

Bouguer's Law (1729). In our discussion of optical phenomena in crystals we assumed that the crystals were perfectly transparent. In actual fact all crystals absorb light transmitted by them to some extent. Let I_0 be the intensity of a parallel beam of light incident normally on a crystal, x_3 the path traversed by the beam in the crystal, and I_{x_3} the intensity transmitted by the crystal. We shall assume that the plate has a unit thickness, i.e., $x_3 = 1$ cm. The transmitted intensity is then given (neglecting reflection) by

$$I_1 = I_0 \alpha,$$

where α is a positive proper fraction, since the intensity of the transmitted light is smaller than the intensity of the incident light. The quantity $\alpha = \dfrac{I_1}{I_0}$ characterizes the transparency of the medium under consideration, and the lower is α the less transparent is the medium. For ideally transparent media $\alpha = 1$ and for perfectly opaque media $\alpha = 0$.

If light transmitted by a 1 cm thick crystal plate is again passed through a similar plate then I_1 will be the incident intensity and I_2 the transmitted intensity. It is natural to assume that the relation between I_2 and I_1 is the same as that between I_1 and I_0 so that we may write

$$I_2 = I_1 \alpha = I_0 \alpha^2.$$

Continuing our analysis in this way we are led to the following general formula

$$I_{x_3} = I_0 \alpha^{x_3}, \tag{125}$$

or

$$\ln \frac{I_{x_3}}{I_0} = x_3 \ln \alpha.$$

174

In an earlier section (p. 68) the quantity $\frac{Ix_3}{I_0}$ was given the name of transmission. We now see that the natural logarithm of the transmission is proportional to the path x_3 traversed by the light. The coefficient of proportionality between these two quantities is $\ln \alpha$. The natural logarithm of the proper fraction α is a negative quantity which we shall denote by $- k$.

Thus

$$\ln \alpha = - k.$$

This means that

$$\alpha = e^{-k}.$$

Substituting this expression for α into (125) we obtain Bouguer's law:

$$I_{x_3} = I_0 e^{-k x_3}. \tag{126}$$

The positive quantity k is known as the absorption coefficient. For a perfectly transparent medium $k = 0$, and for a perfectly opaque medium $k = \infty$.

The absorption coefficient is a function of the wavelength of light in the absorbing medium ($\lambda = \frac{\lambda_0}{n}$). Frequently, this dependence is conveniently indicated by writing

$$k = \frac{4\pi \varkappa}{\lambda}, \tag{127}$$

where $\varkappa$ is the absorption index.

Physical Reasons for the Absorption of Light. The main reason for dispersion and absorption of light is the interaction between vibrations in the incident light and the vibrations of particles (mainly electrons and ions) in the medium which are excited by the incident radiation. Dispersion in its pure form is due to the interaction of light vibrations with undamped vibrations of the particles in the medium, while absorption of light, i.e., the transformation of luminous energy into other forms of energy, is due to the interaction of light vibrations with damped vibrations of particles in the medium. In the absence of resonance between the exciting and excited vibrations one has normal, i.e., not very pronounced, dispersion and absorption. In the presence of resonance, these phenomena assume an anomalous and very pronounced character. It follows that the absorption coefficient should depend both on the nature of the incident light, i.e., on frequency, and the structure of the medium, i.e., on the presence of particles which can vibrate on definite frequencies under the action of the variable electric field of the light waves.

Absorption of light in crystals can also be explained by the fact that any dielectric is also a conductor, and in conducting media the electromagnetic waves induce not only displacement currents but also conduction currents, which leads to the dissipation of heat in crystals. In transparent crystals conductivity is low and has little effect on the optical properties, in absorbing crystals it is appreciable, and in doubly absorbing media it is very anisotropic.

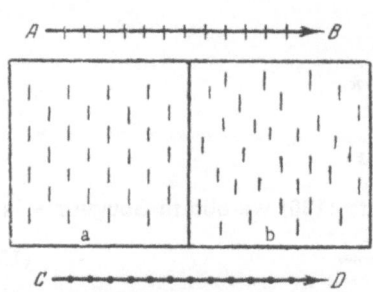

Fig. 158. Linear oscillators in a crystal (a) and a "texture" (b). Polarized ray *AB* is absorbed, while polarized ray *CD* is not absorbed by these media.

Anisotropic Absorption of Light. Absorption of light is observed in gases and liquids and also in crystals. From the point of view of optical crystallography, only those aspects of absorption are of interest where the absorption differs from the corresponding phenomena in amorphous noncrystalline media. This means that one is concerned with anisotropic absorption, i.e., the dependence of the absorption coefficient on the direction of the electric vector of the light wave in the crystal. The existence of anisotropic absorption of light in crystals can be established by direct observation, or it can be predicted on the basis of general ideas on crystal structure. In particular, one may expect that a crystal or "texture" consisting of parallel linear oscillators, say long molecules, should absorb light more strongly when the vibrations are along the length of the oscillators (ray *AB* in Fig. 158) than when the vibrations are perpendicular to the length of the oscillators (ray *CD* in Fig. 158). In other words, absorption is determined not by the direction of propagation but by the direction of light vibrations. This means that two waves of the same frequency which travel along the same direction in a given crystal could, in general, have different absorption coefficients. We shall call this property, namely, the existence of two absorption coefficients for waves of a given frequency and traveling in the same direction, the double absorption of light. The above difference in the values of k for two waves of the same frequency should not be confused with the difference between the values for k for waves of different frequency, which was discussed earlier (cf. (127)).

176

Pleochroism. Unequal absorption of light of different wavelengths leads to the fact that absorbing crystals appear colored when natural white light is transmitted by them. It is easy to understand and to verify experimentally that the hue of a crystal plate, i.e., the relative fraction of the intensity of monochromatic rays of different frequency, should be a function of the thickness of the plate. Clearly, very thin absorbing plates will not be colored since in this case even very different values of k for rays of different frequency will not lead to appreciable absorption. As the thickness of the plate increases, the difference in the values of k for different frequency will play a more and more important role in the coloration of the transmitted light: the hue will become denser and displaced along the spectrum toward the more weakly absorbed rays. This can be investigated quantitatively using formula (126) if the absorption coefficient k is known as a function of wavelength.

Optical crystallography is concerned with the coloration associated with double absorption rather than the coloration which depends on the thickness of the plate and is determined by the different values of k for different frequencies, i.e., the important aspect from the point of view of optical crystallography is the difference in k for waves of a given frequency but different vibration directions. The coloration of crystals due to both of the above causes has received the name of pleochroism.

The phenomenon of pleochroism was first discovered by Cordier in 1809 in a mineral which has subsequently been called cordierite.

Observation of Pleochroism. Pleochroism can, of course, be only observed in uniaxial and biaxial crystals.

If one cuts a cube of diaspore, which is an orthorhombic crystal, in such a way that the faces of the cube are parallel to the symmetry planes of the crystal, then if one observes white light through such a crystal in the direction of its principal axes, three hues can be seen: blue, violet, and green (trichroisn). Under similar conditions, the hexagonal tourmaline will in accordance with its optical symmetry give only three typical hues, namely, green and brown (dichroisn).

Quantitative studies of pleochroism are carried out with parallel polarized beams of light and a single Nicol prism.

1. The pleochroism of uniaxial crystals can be investigated with only one plate cut parallel to the optic axis. If such a plate is rotated until the vibrations of the transmitted polarized light are parallel to

the optic axis of the crystal, the observed hue is the hue of the extraordinary rays. If the plate is then rotated through a further 90° the observed hue is the hue of the ordinary rays. In the intermediate positions of the plate some mean hue is observed which represents a combination of the hues of the ordinary and extraordinary rays. In the section perpendicular to the optic axis, the only hue which can be observed is that of the ordinary rays and is independent of the rotation of the plate about the optic axis of the crystal.

In order to determine the principal coefficients k_1 and k_2 (cf. p. 181), the observations should, of course, be carried out in monochromatic light of a given frequency.

2. In order to investigate the pleochroism of orthorhombic crystals two plates are sufficient and must be cut so that one of them is perpendicular to one of the three principal axes of the crystal and the other is perpendicular to one of the other two principal axes. One can, of course, use a single plate (prism) provided its faces contain two of the above three directions. In order to determine the coefficient k_1, k_2 and k_3 the observations should be carried out in monochromatic light just as in the case of uniaxial crystals.

3. Two plates are also sufficient in a rough investigation of the pleochroism of monoclinic crystals. One of the plates should be cut along a plane of symmetry of the crystal, or perpendicular to its twofold axis. This axis (or normal to a symmetry plane of the crystal) should, in accordance with the symmetry of the crystal, be the k_3 axis in the crystallophysical setting or the k_2 axis in the ordinary crystallographic setting. By rotating the plate about its normal, the principal hues are determined by the maximum contrast between them. It is important to note that the vibration directions corresponding to the two principal hues do not coincide in monoclinic crystals with the principal absorption axes k_1 and k_2 in the crystallophysical setting, or the k_1 and k_3 axes in the usual crystallographic setting. The second plate is cut perpendicular to one of the vibration directions corresponding to the two principal colors. This second plate may be used to find the third principal hue and observe again one of the other two.

4. Investigation of pleochroism of triclinic crystals is complicated by the fact that none of the axes of the absorption indicatrix coincide with the axes of the optical indicatrix describing double refraction. The determination of the three principal hues in triclinic crystals can only be done by trial and error using a sufficiently large number of differently oriented crystal plates. It should be noted that, generally

speaking, the methods for investigating pleochroism of crystals are not well developed.

It is clear from the above discussion that all crystals can be divided into the same five symmetry groups in relation to absorption ($\infty/\infty \cdot m$; $m \cdot \infty : m$; $m \cdot 2 : m$; $2 : m$; $\overline{2}$), as in relation to double refraction (p. 60).

The pleochroism of a polished precious stone may be detected by means of Haidinger's lens (1845) which enables one to see the stone in polarized light in two positions simultaneously. This lens consists of a rhombohedron of Iceland spar S, two glass prisms G, and a lens L (Fig. 159). The crystal under investigation is placed against the aperture in the lens holder.

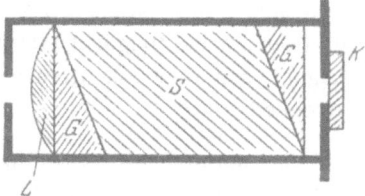

Fig. 159. Haidinger's lens. S) Iceland spar prism; G, G) glass prism; K) crystal plate; L) lens.

In order to detect and investigate qualitatively the pleochroism of a plate cut from a uniaxial crystal in a direction perpendicular to the optic axis, it is convenient to use convergent polarized light and only one Nicol prism (upper or lower) of the polarizing instrument. In this case, the instrument field of view contains four colored sectors (beams) which

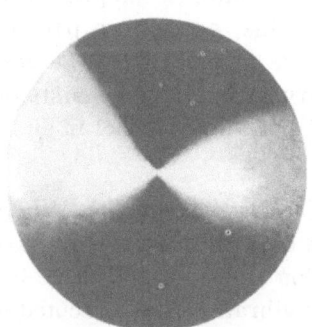

Fig. 160. The appearance of "brushes" in tourmaline.

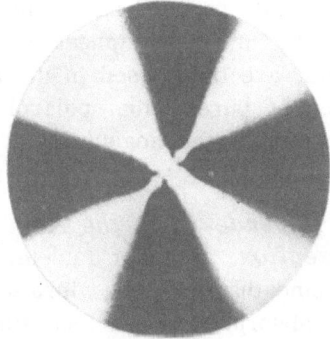

Fig. 161. The appearance of "brushes" in biaxial crystals cut perpendicular to the optic axis.

gradually merge into each other and are such that opposite sectors have the same hue and neighboring sectors have different hues (Fig. 160).

An analogous picture is observed in the case of biaxial crystals cut along a plane perpendicular to the acute bisectrix (Fig. 161). The origin of these pictures is self-explanatory.

In general physics courses, the phenomenon of pleochroism is usually demonstrated using green tourmaline plates cut parallel to the optic axis. If a beam of natural light is passed through such a relatively thick plate, then due to the strong difference in the absorption coefficients for ordinary and extraordinary rays, the emergent light will include the less absorbed extraordinary green rays, while the ordinary rays are almost completely absorbed. The plate therefore behaves in a way similar to a Nicol prism. By using two such plates (tourmaline tongs) it may be demonstrated that they will transmit light when they are parallel, and will extinguish light when they are crossed.

As was pointed out above (p. 28), artificial pleochroic films have been widely used under the name of polaroid, or polarizing light filters, after Land (1932) organized their commercial manufacture. Land prepared his polaroid films from the strongly pleocroic crystalline material quinine iodosulfate ($4Ch \cdot 3H_2SO_4 \cdot 2HI_3 \cdot xH_2O$, where Ch is the quinine molecule $C_{20}H_{24}N_2O_2$). The optical properties of crystals of this substance were known by Herapath fifty years before, who used this substance to grow monocrystal plates.

The classical polaroid consists of a transparent plastic film in which submicroscopic needle-like crystals are uniformly oriented, the substance used being known as herapathite. At the present time, various other substances containing iodine, as well as pure iodine itself, are being used in the manufacture of polaroids. The preparation of pure iodine polaroids is very simple and consists of two operations: the soaking of a film of polyvinyl alcohol in an iodine solution and its subsequent stretching.

Dependence of the Absorption Coefficient on the Vibration Direction. In order to determine this dependence it is necessary to assume that for each vibration direction in the crystal there is only one absorption coefficient, since linear vibrations are executed symmetrically with respect to a given straight line. Next, the absorption coefficient is by definition an essentially positive quantity, and absorption coefficients cannot differ from each other in being right-handed or left-handed, i.e., left-handed and right-handed absorption coefficients do not exist. It follows that since the absorption coefficient is undoubtedly a directed quantity and is neither a polar nor an axial vector, it must be a tensor, and moreover, a polar tensor. Thus the absorption coefficient can be considered, in the first approximation, as a quadratic function of the direction cosines c_1, c_2, c_3 (Voigt 1884 – 1902) by analogy with dielectric constant, expansion coefficient, specific rotation of the plane of polarization, etc.

In a specially chosen principal set of coordinates this function has the form (cf. (98))

$$k = k_1 c_1^2 + k_2 c_2^2 + k_3 c_3^2. \tag{128}$$

The variable quantity k can stand for either the absorption coefficient referred to the vibration direction in the ray (k_S), or the absorption coefficient referred to the vibration direction in the wave normal (k_N). The difference between these almost equal quantities will be neglected. The constants k_1, k_2, k_3 are called the principal absorption coefficients of the crystal.

The surface whose radius vectors are proportional to k, is known as the absorption coefficient surface.

Let us consider the form of equation (130) and of the corresponding surface for different ratios of the absorption coefficients.

For positive values of k_1, k_2, k_3, the following cases are possible.

1. All three coefficients are equal

$$k = k_3 (c_1^2 + c_2^2 + c_3^2) = k_3. \tag{129}$$

The absorption coefficient surface is then a sphere. Only optically isotropic bodies such as cubic crystals and isotropic amorphous bodies (glass) can have such an absorption surface. The symmetry of the surface is $\infty / \infty \cdot m$. For waves of different frequency the sphere has a different radius k_3.

2. Two coefficients equal ($k_1 = k_2$). The third coefficient which is different from the first two can be either zero or not.

a) When $k_3 \neq 0$ we have

$$k = k_1 (c_1^2 + c_2^2) + k_3 c_3^2. \tag{130}$$

The absorption coefficient surface is now an ovaloid-like surface of revolution which we have met already (Figs. 118 and 119), and can be either prolate or oblate. Only uniaxial crystals have such a surface and its symmetry is $m \cdot \infty : m$.

For waves of different frequency the surface has a different axial ratio $k_1 : k_3$.

b) When $k_3 \neq 0$ and $k_1 = k_2 = 0$ we have

$$k = k_3 c_3^2. \tag{131}$$

the corresponding surface of revolution, which has a $m \cdot \infty : m$ symmetry, was also described before (Fig. 121).

We know that there are no crystals which are perfectly transparent even for special vibration directions in the crystal (in the present case, for all the directions across the optic axis of the uniaxial crystal). It follows that the above surface describes the absorption of light only approximately, and only for waves of a given wavelength. With this reservation, the latter surface can enter the family of surfaces (130) only as a special case.

c) When $k_3 = 0$ and $k_1 = k_2 \neq 0$ we have

$$k = k_1 (c_1^2 + c_2^2). \tag{132}$$

The corresponding surface of revolution was also mentioned above (Fig. 122) and its symmetry is $m \cdot \infty : m$.

This surface, similarly to the previous surface, can be found in the family of surfaces of a uniaxial crystal (130) only as a special case for a wave of a strictly defined frequency.

3. All the coefficients are different: $k_1 \neq k_2 \neq k_3$, but one of them, for example k_3, can either be finite or zero.

a) When $k_3 \neq 0$ we have

$$k = k_1 c_1^2 + k_2 c_2^2 + k_3 c_3^2. \tag{133}$$

The surface described by this equation was also mentioned earlier (Fig. 125). It corresponds to the general case of absorption of light by biaxial crystals and its symmetry is $m \cdot 2 : m$. The ratio $k_1 : k_2 : k_3$ is different for different frequencies.

In an orthorhombic crystal all the surfaces belonging to the family (133) have their symmetry elements along the corresponding symmetry elements of the crystal.

In a monoclinic crystal, one of the symmetry axes of each surface of this family has a fixed position and is parallel to the twofold axis of the crystal, or the normal to its symmetry plane.

In triclinic crystals, each "monochromatic" surface has an indeterminate orientation.

b) When $k_3 = 0$ we have

$$\overset{,}{k} = k_1 c_1^2 + k_2 c_2^2. \tag{134}$$

The corresponding surface was shown earlier in Fig. 127. It can occur in biaxial crystals as a degenerate surface only for waves of a strictly defined frequency.

Let us now supplement the above discussion by summarizing the similarities and differences between the absorption coefficient surfaces and the gyration surfaces. In spite of the fact that all these surfaces are described by the same quadratic function, the symmetry groups of the gyration surfaces are the acentric groups $2 : 2$, $\bar{4} \cdot m$, $\infty : 2$, ∞ / ∞, while the symmetry groups of the absorption surfaces are the centrally symmetric groups $m \cdot 2 : m$, $m \cdot \infty : m$, $\infty / \infty \ m$. The radius vectors of the gyration surfaces can be looked upon as segments of a straight line twisted along a right-handed or left-handed screw, while the radius vectors of absorption surfaces may be looked upon as untwisted segments of straight lines.

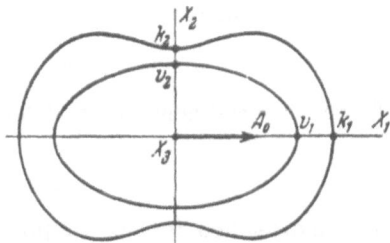

Fig. 162. Concerning the derivation of formula (135).

The basic difference between a k-surface and a ρ-surface is that the radius vectors of the first coincide with the vibration direction, while the radius vectors of the second coincide with the ray direction (or the direction of the wave normals). This means that a gyration surface can be used to determine directly the rotation of the plane of polarization for any given ray (or wave normal), while the absorption coefficient for a given ray (or wave normal) can only be obtained from the absorption coefficient surface if the corresponding vibration direction is also known.

Equation of a Plane Wave in a Doubly Absorbing Crystal.

Consider a linearly polarized plane wave having an amplitude A_0 and incident on a plate of an orthorhombic crystal cut along a plane perpendicular to one of the principal axes, for example, the X_3 axis (Fig. 162). Such a plate should cut the Fresnel ellipsoid in an ellipse with semiaxes v_1 and v_2, and the absorption coefficient surface in an ovaloid-like curve with semiaxes k_1, k_2. In accordance with the symmetry conditions, the v_1 and k_1 axes should coincide with the X_1 axis. We shall consider the case where the vibration direction in the incident wave also coincides with the X_1 axis.

The intensity I_{x_1} of the incident wave should in this case fall off according to (126), i.e., it is given by

$$I_{x_1} = I_0 e^{-k_1 k_1}, \tag{135}$$

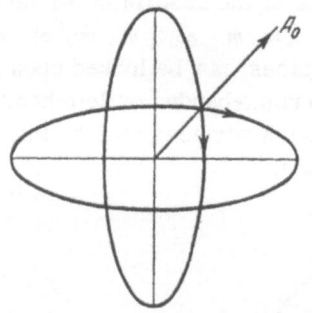

where I_0 is the initial intensity of the wave. In addition to the decrease in the intensity there should also be a decrease in the amplitude A_0. Let A_{x_3} denote the amplitude at a depth x_3 within the crystal. Since the intensity is proportional to the square of the amplitude we have

$$I_{x_1} = A_{x_3}^2 \text{ and } I_0 = A_0^2,$$

Fig. 163. Elliptical double refraction in absorbing crystals.

and hence using (135)

$$A_{x_3} = A_0 e^{\frac{-k_1 x_3}{2}}. \tag{136}$$

Substituting the last expression for the amplitude into one of the earlier equations for a plane wave, for example equation (28), and remembering that in our case the wave should, in accordance with the Fresnel construction, travel with the velocity v_1 we have

$$x_1 = A_0 e^{\frac{-k_1 x_3}{2}} \sin\left[\frac{2\pi}{T}\left(t - \frac{x_3}{v_1}\right)\right]. \tag{137}$$

If the vibration direction in the incident wave is at an angle α to the X_1 axis, then in accordance with Malus's law (p. 62) the wave should decompose into two waves with amplitudes $A_0 \cos \alpha$ and $A_0 \sin \alpha$. The first wave will be propagated in the crystal with a velocity v_1 and an absorption coefficient k_1, and the second wave will have a velocity v_2 and an absorption coefficient k_2. On leaving the crystal, the two waves will recombine into a single wave with elliptical vibrations.

We have considered one of the simplest cases of transmission of a light wave through a crystal. In the general case, when the plates are cut obliquely from a biaxial crystal, the equation for the wave is complicated by the fact that the principal axes of the section of the Fresnel ellipsoid do not coincide with the principal axes of the section through the absorption coefficient surface. A detailed study of this problem leads to the conclusion that in this case a linearly polarized incident wave is decomposed within the crystal plate (in the diagonal position) into two waves which travel with different velocities, and whose vibrations are over identical right-handed or left-handed crossed ellipses (Fig. 163). In other words, doubly absorbing crystals give rise to elliptical double refraction which differs from the elliptical double refraction described above in that in this new form the end point of the electrical vector moves in the same direction in both ellipses, i.e., they are both right-handed or both left-handed; in the previous form of elliptical double refraction one of the ellipses was right-handed and the other left-handed.

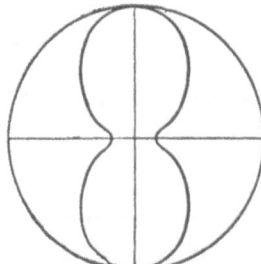

Fig. 164. Absorption surface of two sheets for tourmaline.

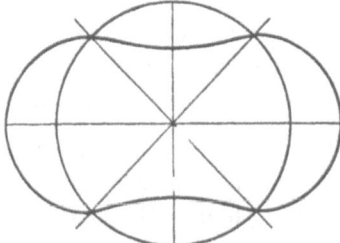

Fig. 165. Absorption surface of two sheets for a crystal of potassium and cobalt sulfate.

Absorption Ovaloid and Absorption Surface of Two Sheets. By absorption ovaloid we understand an ovaloid whose principal axes are equal to k_1, k_2, k_3. The equation of such an ovaloid is of the form

$$k_N^2 = k_1^2 c_1^2 + k_2^2 c_2^2 + k_3^2 c_3^2. \qquad (138)$$

The role played by the ovaloid in double absorption of light is similar to that played by the velocity ovaloid (47) in double refraction. In particular, the absorption ovaloid may be used to construct an absorption surface of two sheets in a similar way to that used to obtain the wave surface of two sheets from the Fresnel ellipsoid (Fig. 35). The radius vectors of the absorption surface of two sheets are seg-

185

ments of straight lines equal in length to the absorption coefficients k_N of the crystal in the direction of the wave normals, and not the coefficients k in the vibration directions, as was the case in the surface described by equation (128). This means that for each direction in the crystal there are, in general, two waves having different absorption coefficients k'_N and k''_N.

When $k_1 = k_2 = k_3$, the absorption ovaloid becomes a sphere. It describes absorption of light in cubic crystals.

When $k_1 = k_2 \neq k_3$ the absorption ovaloid is a surface of revolution which has the same symmetry as an ellipsoid of revolution. The corresponding surface of two sheets has the same symmetry. These surfaces describe double absorption in uniaxial crystals.

When $k_1 \neq k_2 \neq k_3$ we have a general ovaloid. This ovaloid, and the corresponding surface of two sheets, describe double absorption in biaxial crystals. Both surfaces have a $m \cdot 2 : m$ symmetry.

The general ovaloid has two circular sections but the planes of these sections do not, in general, coincide with the planes of the corresponding sections through the optical indicatrix. It follows that in a biaxial crystal there are two directions in which double absorption is absent and two directions in which double refraction is absent.

The above theoretical discussion is on the whole confirmed by measurements of absorption coefficients of moderately absorbing crystals. Figures 164 and 165 show experimental absorption surfaces of two sheets for tourmaline and potassium sulfate and cobalt sulfate crystals. The theory is not sufficiently well developed in the case of strongly absorbing crystals, for example, crystals of almost opaque metals.